T0281032

Synthesis Lectures on Communications

Series Editor

William H. Tranter, Virginia Tech, Blacksburg, VA, USA

This series of short books cover a wide array of topics, current issues, and advances in key areas of wireless, optical, and wired communications. The series also focuses on fundamentals and tutorial surveys to enhance an understanding of communication theory and applications for engineers.

Jerry D. Gibson

Digital Communications

Introduction to Communication Systems

 Springer

Jerry D. Gibson
Department of Electrical and Computer
Engineering
University of California
Santa Barbara, CA, USA

ISSN 1932-1244   ISSN 1932-1708   (electronic)
Synthesis Lectures on Communications
ISBN 978-3-031-19590-7   ISBN 978-3-031-19588-4   (eBook)
https://doi.org/10.1007/978-3-031-19588-4

© The Editor(s) (if applicable) and The Author(s), under exclusive license to Springer Nature Switzerland AG
2023
This work is subject to copyright. All rights are solely and exclusively licensed by the Publisher, whether the whole
or part of the material is concerned, specifically the rights of translation, reprinting, reuse of illustrations, recitation,
broadcasting, reproduction on microfilms or in any other physical way, and transmission or information storage
and retrieval, electronic adaptation, computer software, or by similar or dissimilar methodology now known or
hereafter developed.
The use of general descriptive names, registered names, trademarks, service marks, etc. in this publication does
not imply, even in the absence of a specific statement, that such names are exempt from the relevant protective
laws and regulations and therefore free for general use.
The publisher, the authors, and the editors are safe to assume that the advice and information in this book are
believed to be true and accurate at the date of publication. Neither the publisher nor the authors or the editors give
a warranty, expressed or implied, with respect to the material contained herein or for any errors or omissions that
may have been made. The publisher remains neutral with regard to jurisdictional claims in published maps and
institutional affiliations.

This Springer imprint is published by the registered company Springer Nature Switzerland AG
The registered company address is: Gewerbestrasse 11, 6330 Cham, Switzerland

Contents

About the Author

Jerry D. Gibson is Professor of Electrical and Computer Engineering at the University of California, Santa Barbara. He is Co-author of the books *Digital Compression for Multimedia* (Morgan-Kaufmann, 1998) and *Introduction to Nonparametric Detection with Applications* (Academic Press, 1975 and IEEE Press, 1995) and Author of the textbook, *Principles of Digital and Analog Communications* (Prentice-Hall, second ed., 1993). He is Editor-in-Chief of *The Mobile Communications Handbook* (CRC Press, 3rd ed., 2012), Editor-in-Chief of *The Communications Handbook* (CRC Press, 2nd ed., 2002), and Editor of the book, *Multimedia Communications: Directions and Innovations* (Academic Press, 2000). His most recent books are *Rate Distortion Bounds for Voice and Video* (Coauthor with Jing Hu, NOW Publishers, 2014) and *Information Theory and Rate Distortion Theory for Communications and Compression* (Morgan-Claypool, 2014).

He was Associate Editor for Speech Processing for the *IEEE Transactions on Communications* from 1981 to 1985 and Associate Editor for Communications for the *IEEE Transactions on Information Theory* from 1988–1991. He was IEEE Communications Society Distinguished Lecturer for 2007–2008.

In 1990, he received The Fredrick Emmons Terman Award from the American Society for Engineering Education, and in 1992, he was elected Fellow of IEEE. He was the recipient of the 1993 IEEE Signal Processing Society Senior Paper Award for the Speech Processing area. He received the *IEEE Transactions on Multimedia* Best Paper Award in 2010 and the IEEE Technical Committee on Wireless Communications Recognition Award for contributions in the area of wireless communications systems and networks in 2009.

Data Transmission

1

1.1   Introduction

This chapter is concerned with transmitting discrete-amplitude (digital) signals over bandwidth-constrained communications channels. To achieve the highest possible transmission rate or data rate in bits per second while minimizing the number of errors in received bits, called the *bit error rate* (BER), requires considerable knowledge, skill, and ingenuity. As a result, this chapter covers what seems to be a wide variety of topics. However, each of these topics plays a fundamental role in the design and analysis of data communication systems.

   In previous books, we have represented digital signals by rectangular (time-domain) pulses of a given amplitude and width. The channel bandwidth limitation precludes the use of rectangular pulses here, since the frequency content of a rectangular pulse has a $\sin x/x$ shape. The tails of the $\sin x/x$ function decay very slowly, and hence a rectangular pulse requires a wide bandwidth for undistorted transmission. In Sect. 1.2, we introduce the concept of pulse shaping for restricted bandwidth applications while maintaining zero intersymbol interference with adjacent pulses. Nyquist has shown that there is an upper limit on the rate at which pulses can be transmitted over a bandlimited channel while maintaining zero intersymbol interference. The use of controlled intersymbol interference, called partial response signaling, to achieve the maximum data rate possible in the minimum bandwidth is developed in Sect. 1.3. For many communications channels, and the telephone channel in particular, impairments are present which are deterministic in that they can be measured for a given transmission path and they do not change perceptibly over a short time period. This deterministic distortion must be well understood, and it is the subject of Sect. 1.4. A good way to get a subjective idea of the distortion contributed by a transmission path for a particular data transmission system is to display what is called the system eye pattern. Eye patterns are discussed in Sect. 1.5, followed

© The Author(s), under exclusive license to Springer Nature Switzerland AG 2023
J. D. Gibson, *Digital Communications*, Synthesis Lectures on Communications,
https://doi.org/10.1007/978-3-031-19588-4_1

in Sect. 1.6, by a development of channel equalization techniques, which attempt to correct any deterministic distortion contributed by the channel. In Sect. 1.7, we develop data scramblers for randomizing the input message sequence and which thus aid the performance of automatic, adaptive equalizers. The difficult problems of carrier acquisition and symbol synchronization are considered briefly in Sect. 1.8, followed by Sect. 1.9, which unifies the chapter by discussing example modem designs.

1.2   Baseband Pulse Shaping

As noted in the preceding section, we have always represented digital signals in earlier chapters by a rectangular time-domain pulse. This pulse shape is convenient in that it is wholly contained within its allocated interval and hence does not interfere with pulses in adjacent intervals. The rectangular time-domain pulse does require excessive bandwidth, however. This is easily seen by considering a unit amplitude, rectangular pulse of width τ centered at the origin,

$$p(t) = \begin{cases} 1, & |t| \leq \frac{\tau}{2} \\ 0, & \text{otherwise,} \end{cases} \qquad (1.2.1)$$

which has the Fourier transform

$$P(\omega) = \mathcal{F}\{p(t)\} = \tau \frac{\sin(\omega\tau/2)}{\omega\tau/2}. \qquad (1.2.2)$$

Since the zero crossings of $P(\omega)$ occur at $\omega = \pm 2n\pi/\tau$, the narrower the pulse, the wider the bandwidth required. Furthermore, the tails of $\sin(\omega\tau/2)/(\omega\tau/2)$ decay very slowly for increasing ω. As a result, the rectangular time-domain pulse is not acceptable for bandwidth-constrained applications.

To begin our search for suitable pulse shapes, we must be more specific concerning our requirements. First, we want the chosen pulse shape to have zero intersymbol interference at the pulse sampling times. Since, at the receiver, pulses are usually sampled at the exact center of their allocated interval, this means that the desired pulse shape is zero at the center of all pulse intervals other than its own. Of course, the rectangular pulse satisfies this requirement, since it is identically zero outside its own interval. However, we are being less restrictive here in that the time-domain pulse may be nonzero in other pulse intervals as long as it is zero at the center of every other pulse interval (the pulse sampling times). Second, the desired pulse shape must have a Fourier transform that is zero for ω greater than some value, say ω_m. This requirement satisfies our bandwidth constraint. Third, we would prefer that the tails of the time-domain pulse shape decay as rapidly as possible, so that jitter in the pulse sequence or sampling times will not cause significant intersymbol interference. This last requirement is a very pragmatic one, but it is necessary, since any data transmission system will have some timing jitter, which will cause pulse sampling to occur at times other than the exact center of each pulse interval. If the pulse tails are not sufficiently damped, sampling in the presence of timing jitter can cause substantial intersymbol interference.

A good starting point in our search for a useful pulse shape is the requirement that its Fourier transform be identically zero for $\omega > \omega_m$. By the symmetry property of Fourier transforms, we know that a sin x/x time-domain pulse has a rectangular frequency content. In particular, the pulse shape

$$p(t) = \frac{\sin(\pi t/\tau)}{\pi t/\tau} \tag{1.2.3}$$

has the Fourier transform

$$P(\omega) = \begin{cases} \tau, & |\omega| \leq \frac{\pi}{\tau} \\ 0, & \text{otherwise} \end{cases} \tag{1.2.4}$$

Not only does $p(t)$ in Eq. (1.2.3) have a restricted frequency content, but it also is identically zero at the points $t = \pm n\tau$ about its center. Thus, if the pulses are spaced τ seconds apart, this $p(t)$ achieves zero intersymbol interference at the pulse sampling times. Unfortunately, our third requirement on the tails of the pulse is the demise of $p(t)$ in Eq. (1.2.3). The tails of $p(t)$ only decay as $1/t$, and it can be shown that for very long messages even a small error in the sampling times produces substantial intersymbol interference. In fact, even a small error in sampling times can cause the tails of a very long pulse sequence to sum as a divergent series (see Problem 1.3).

Since $p(t)$ in Eq. (1.2.3) satisfies two of our three requirements, we are led to inquire whether this $p(t)$ can be modified to have faster-decaying tails while retaining the same zero crossings and bandlimited feature. Fortunately, Nyquist (1928) has derived a requirement on the Fourier transforms of possible pulse shapes which, if satisfied, guarantees that the zero crossings of $p(t)$ in Eq. (1.2.3) are retained. However, the bandwidth is increased and there may be additional zero crossings. This criterion (called *Nyquist's first criterion*) is that the Fourier transform $G(\omega)$ of a real and even pulse shape $g(t)$ satisfies

$$G\left(\frac{\pi}{\tau} - x\right) + G\left(\frac{\pi}{\tau} + x\right) = \text{constant} \tag{1.2.5}$$

for $0 \leq x \leq \pi/\tau$, where τ is the spacing between pulse sampling instants. The proof of this result can be obtained by allowing the bandwidth to increase over that in Eq. (1.2.4), so that $2\pi B > \pi/\tau$, and writing

$$p(t) = \frac{1}{2\pi} \int_{-2\pi B}^{2\pi B} P(\omega) e^{j\omega t} d\omega.$$

We want zero intersymbol interference at the sampling instants, so we let $t = n\tau$,

$$p(n\tau) = \frac{1}{2\pi} \int_{-2\pi B}^{2\pi B} P(\omega)e^{j\omega n\tau} d\omega.$$

For $2B\tau = \text{integer} = K > 1$ and since $B > 1/2\tau$,

$$p(n\tau) = \sum_{k=-K}^{K} \frac{1}{2\pi} \int_{(2k-1)\pi/\tau}^{(2k+1)\pi/\tau} P(\omega)e^{j\omega n\tau} d\omega$$

$$= \sum_{k=-K}^{K} \frac{1}{2\pi} \int_{-\pi/\tau}^{\pi/\tau} P\left(\lambda + \frac{2k\pi}{\tau}\right) e^{jn\tau(\lambda + 2k\pi/\tau)} d\lambda$$

$$= \frac{1}{2\pi} \int_{-\pi/\tau}^{\pi/\tau} \sum_{k=-K}^{K} P\left(\lambda + \frac{2k\pi}{\tau}\right) e^{jn\tau\lambda} d\lambda.$$

For zero intersymbol interference at the sampling times, $p(0) = 1$, $p(n\tau) = 0$, $n \neq 0$, which requires that

$$\sum_{k=-K}^{K} P\left(\omega + \frac{2k\pi}{\tau}\right) = \begin{cases} \tau, & |\omega| \leq \frac{\pi}{\tau} \\ 0, & |\omega| > \frac{\pi}{\tau}. \end{cases} \tag{1.2.6}$$

Equation (1.2.6) implies the criterion in Eq. (1.2.5).

Pulse shapes with Fourier transforms that satisfy Eqs. (1.2.5) and (1.2.6) are not guaranteed to have rapidly decaying tails nor are they guaranteed to be bandlimited; however, they are guaranteed to have zero crossings at the nulls of $p(t)$ in Eq. (1.2.3). Thus, by investigating the class of bandlimited Fourier transforms that satisfy Eqs. (1.2.5) and (1.2.6), we may be able to find pulse shapes that satisfy our third requirement as well.

An important pulse shape that satisfies Nyquist's first criterion and our three requirements is called the *raised cosine pulse,* and for real and even pulse shapes $p(t)$, it has a Fourier transform that is even about $\omega = 0$ and that for positive ω is given by

$$P(\omega) = \begin{cases} \tau, & 0 \leq \omega \leq \frac{\pi(1-\alpha)}{\tau} \\ \frac{\tau[1+\cos\{\tau[\omega - \pi(1-\alpha)/\tau]/2\alpha\}]}{2}, & \frac{\pi(1-\alpha)}{\tau} \leq \omega \leq \frac{\pi(1+\alpha)}{\tau} \\ 0, & \omega > \frac{\pi(1+\alpha)}{\tau}. \end{cases} \tag{1.2.7}$$

The impulse response corresponding to Eq. (1.2.7) is

$$p(t) = \frac{\sin(\pi t/\tau)}{\pi t/\tau} \cdot \frac{\cos(\alpha\pi t/\tau)}{1 - (4\alpha^2 t^2)/\tau^2}. \tag{1.2.8}$$

It is easy to see that this pulse shape satisfies our three requirements. From Eq. (1.2.7), $p(t)$ is bandlimited, while from Eq. (1.2.8) we observe that it has zero crossings at least at $\pm n\tau$ and its tails decay as $1/t^3$.

A pulse $p(t)$ given by Eq. (1.2.8) or a Fourier transform $P(\omega)$ as shown in Eq. (1.2.7) is said to have *raised cosine spectral shaping with* $100\alpha\%$ *cosine roll-off.* Sketches of $p(t)$ and $P(\omega)$ for various α are shown in Fig. 1.1a, b, respectively. It is important to note that although $p(t)$ is called a raised cosine pulse, the adjective *raised cosine* refers to the frequency-domain shaping, not the time-domain shaping. Two equivalent forms of Eq. (1.2.7) often appear in the literature,

$$P(\omega) = \begin{cases} \tau, & 0 \leq \omega \leq \frac{\pi(1-\alpha)}{\tau} \\ \tau \cos^2\left\{\tau\left[\omega - \frac{\pi(1-\alpha)}{\tau}\right]/4\alpha\right\}, & \frac{\pi(1-\alpha)}{\tau} \leq \omega \leq \frac{\pi(1+\alpha)}{\tau} \\ 0, & \omega > \frac{\pi(1+\alpha)}{\tau} \end{cases} \tag{1.2.9}$$

and

$$P(\omega) = \begin{cases} \tau, & 0 \leq \omega \leq \frac{\pi(1-\alpha)}{\tau} \\ \frac{\tau[1-\sin\{\tau(\omega-\pi/\tau)/2\alpha\}]}{2}, & \frac{\pi(1-\alpha)}{\tau} \leq \omega \leq \frac{\pi(1+\alpha)}{\tau} \\ 0, & \omega > \frac{\pi(1+\alpha)}{\tau}. \end{cases} \tag{1.2.10}$$

By inspection of Fig. 1.1, we see that the parameter α is extremely important, since it affects both the required bandwidth and the "size" of the tails of $p(t)$. For $\alpha = 1$, the tails of $p(t)$ are highly damped, but the required bandwidth is at the maximum value $\omega_m = 2\pi/\tau$. As α is decreased toward zero, the required bandwidth decreases, but the tails of $p(t)$ become increasingly significant. The usual range of α for data transmission is $0.3 \leq \alpha \leq 1.0$, with α as low as 0.1 for some sophisticated systems.

When raised cosine pulses are used for data transmission, note that the pulses occur every τ seconds; hence the pulse rate, *symbol rate,* or *baud rate,* as it is usually called, is $1/\tau$ symbols/s. If the transmitted pulses take on only two possible levels, each pulse represents 1 bit, and the bit rate = symbol rate = $1/\tau$ bits/s.

Example 1.2.1 It is desired to send data at a rate of 1000 bits/s using a positive raised cosine pulse with 100% roll-off to represent a l and no pulse to represent a 0.

(a) What is the required symbol rate?

(b) What is the bandwidth of a transmitted positive pulse?

(c) If we wish to send the sequence 1011001, sketch the transmitted pulse stream.

Solution

(a) Since we are using two-level pulses, symbol rate = bit rate = 1000 symbols/s.

(b) Since $\alpha = 1$, the bandwidth is $2\pi/\tau$, where τ = pulse spacing and $1/\tau$ = symbol rate = 1000 symbols/s. Thus the pulse bandwidth is $2\pi/\tau = 2000\pi$ rad/s.

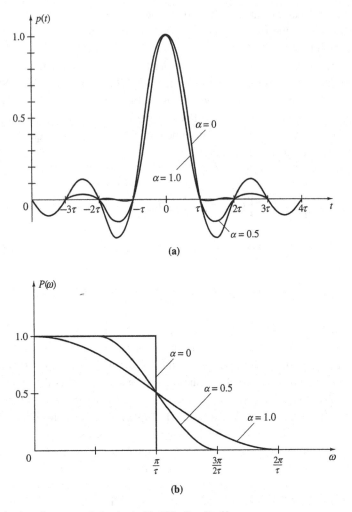

Fig. 1.1 Raised cosine spectral shaping with $100\alpha\%$ roll-off

(c) The transmitted pulse stream is sketched in Fig. 1.2. To emphasize their individual
contributions, the pulses are not added point by point as they would appear on an
oscilloscope.

Perhaps we should note at this point that if we wish to determine the bandwidth of a
transmitted pulse stream with a given pulse shaping, in general, we need only consider
the Fourier transform of the time-domain pulse shape. To show this in the most general situation, we need some expertise in random processes. Deferring this more detailed

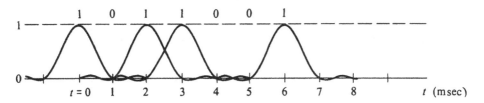

Fig. 1.2 Transmitted pulse stream, Example 1.2.1

development, we can still get an idea of what is happening by examining the special case of a periodic sequence of alternating "pulse" and "no pulse". In this case we know from our knowledge of Fourier series that the spacing of the spectral lines depends on the period but that the envelope of the spectral lines depends on the particular pulse shape. We shall thus use the Fourier transform of the specific time-domain pulse being employed as an indicator of data stream bandwidth.

A standard question that arises in data transmission studies is the following. Given a bandlimited channel of bandwidth $2\pi W$ rad/s, what is the maximum number of noninterfering pulses that can be transmitted per second over this channel? By reconsidering the raised cosine spectral shaping at $\alpha = 0$, we can surmise a possible answer to this question. At $\alpha = 0$, we obtain as our pulse shape $p(t)$ in Eq. (1.2.3) with Fourier transform $P(\omega)$ in Eq. (1.2.4). Thus it seems that the minimum bandwidth within which we can send $1/\tau$ pulses/s is $\pi/\tau = 2\pi W$ rad/s or $W = 1/2\tau$ Hz. We are led to guess then that in a bandwidth of W Hz, we can transmit a maximum of $2W$ pulses/s. We have therefore obtained a special case of Nyquist's (1924) general result that in a bandwidth of W Hz, a maximum of kW pulses/s can be transmitted without intersymbol interference, where $k \leq 2$ is a proportionality factor depending on the pulse shape and the bandwidth.

Nyquist (1928) also developed two other criteria relating to pulse shaping in data transmission. *Nyquist's second criterion* concerns the required spectral shaping to achieve zero intersymbol interference at the instants halfway between adjacent sampling times, •that is, zero intersymbol interference at the edges of the intervals allocated to individual pulses. For real Fourier transforms, the required spectral content is

$$P(\omega) = \begin{cases} \tau \cos \frac{\omega\tau}{2}, & 0 \leq |\omega| \leq \frac{\pi}{\tau} \\ 0, & |\omega| > \frac{\pi}{\tau}. \end{cases} \tag{1.2.11}$$

The impulse response corresponding to $P(\omega)$ in Eq. (1.2.11) is

$$p(t) = \frac{2}{\pi} \cdot \frac{\cos(\pi t/\tau)}{1 - \left(4t^2/\tau^2\right)}. \tag{1.2.12}$$

The impulse response in Eq. (1.2.12) is zero at all odd positive or negative multiples of $\tau/2$ except for $t = \pm\tau/2$, where $p(t)$ is 1/2. Therefore, if we apply an infinite series of impulses spaced τ seconds apart, denoted $\delta_\tau(t)$, to the input of a filter with the transfer function shown in Eq. (1.2.11), measurements of the output taken halfway between adjacent pairs of impulses will be 1/2 times the sum of the two adjacent impulse weights. The output at these time instants due to other impulses will be zero.

For a binary unit-amplitude "pulse" or "no pulse" data stream, the possible outputs at these time instants will be 0, ½, and +1. For a binary sequence of unit amplitude positive and negative pulses, the possible outputs are +2, 0, and −2. As a result, simple threshold detectors at these time instants can detect transitions between logic 1's and logic 0's.

$P(\omega)$ in Eq. (1.2.11) is one example of an entire class of spectral shapes that preserves the spacing of these transitions. Members of this class can be obtained by adding any real spectral shaping function $[P_a(\omega)]$ with even symmetry about π/τ to $P(\omega)$. That is, any real function $P_a(\omega)$ that satisfies

$$P_a\left(\frac{\pi}{\tau} - x\right) = P_a\left(\frac{\pi}{\tau} + x\right), \tag{1.2.13}$$

$0 \le x \le \pi/\tau$, preserves these transitions. An imaginary function is required to have odd symmetry about π/τ to preserve the transition times.

Example 1.2.2 A 100% roll-off raised cosine pulse satisfies Nyquist's second criterion as well as Nyquist's first criterion. To demonstrate that it satisfies Nyquist's second criterion, we need only show that the difference between a 100% roll-off raised cosine spectrum and $P(\omega)$ in Eq. (1.2.11) satisfies Eq. (1.2.13). Letting $\alpha = 1$ in Eq. (1.2.7) and subtracting Eq. (1.2.11), we obtain

$$P_a(\omega) = \begin{cases} \frac{\tau[1-\cos(\omega\tau/2)]}{2}, & 0 \le |\omega| \le \frac{\pi}{\tau} \\ \frac{\tau[1+\cos(\omega\tau/2)]}{2}, & \frac{\pi}{\tau} \le |\omega| \le \frac{2\pi}{\tau}. \end{cases} \tag{1.2.14}$$

For $0 \le \omega \le \pi/\tau$, we replace ω with $\pi/\tau - x$, and manipulate

$$\frac{\tau[1 - \cos(\pi/\tau - x)\tau/2]}{2} = \frac{\tau[1 - \cos(\pi/2 - x\tau/2)]}{2}$$

$$= \frac{\tau[1 - \sin(x\tau/2)]}{2}. \tag{1.2.15}$$

For $\pi/\tau \le \omega \le 2\pi/\tau$, we have (letting $\omega = \pi/\tau + x$)

$$\frac{\tau[1 + \cos(\pi/\tau + x)\tau/2]}{2} = \frac{\tau[1 + \cos(\pi/2 + x\tau/2)]}{2}$$

$$= \frac{\tau[1 - \sin(x\tau/2)]}{2}, \tag{1.2.16}$$

which upon comparing Eqs. (1.2.15) and (1.2.16) verifies the property.

The verification that the raised cosine pulse satisfies Nyquist's first criterion is left to the problems.

Nyquist's third criterion involves the area under the received waveform during a symbol time interval. In particular, Nyquist specified a spectral shaping function whose impulse response has an area in its interval proportional to the applied impulse weight and zero area for every other symbol interval. This criterion is developed further in Problem 1.11.

1.3 Partial Response Signaling

In Sect. 1.2 we stated that it is not possible to send pulses at a rate of $1/\tau$ pulses per second in the minimum bandwidth of $1/2\tau$ Hz without risking substantial intersymbol interference in the presence of timing jitter. The $100\alpha\%$ roll-off raised cosine pulse shaping in Sect. 1.2 is based on the assumption that the transmitted pulses are independent. Lender (1963, 1964, 1966) showed that by introducing a known dependence or correlation between successive pulse amplitudes, the maximum data rate $1/\tau$ symbols/s could be achieved in the minimum bandwidth of $1/2\tau$ Hz without excessive sensitivity to timing jitter. The penalty for this performance improvement is that more voltage levels need to be transmitted over the channel (However, as noted in Sect. 2.7, this need not result in performance loss for a properly designed receiver.). Lender called his method *duobinary*, while Kretzmer (1965, 1966) introduced the term *partial response* in his papers extending Lender's work. Partial response methods have also been called *correlative techniques* or *correlative coding* (Pasupathy 1977; Lender 1981). In this section we emphasize the two most popular correlative coding schemes, duobinary and class 4 partial response.

For duobinary or class 1 partial response systems, the spectral shaping filter is

$$P(\omega) = \begin{cases} 2\tau e^{-j\omega\tau/2} \cos \frac{\omega\tau}{2}, & |\omega| \leq \frac{\pi}{\tau} \\ 0, & |\omega| > \frac{\pi}{\tau}, \end{cases} \tag{1.3.1}$$

with impulse response

$$p(t) = \frac{4}{\pi} \cdot \frac{\cos[\pi(t - \tau/2)/\tau]}{1 - 4(t - \tau/2)^2/\tau^2}. \tag{1.3.2}$$

The impulse response, $p(t)$, is sketched in Fig. 1.3. If the sampling instant for the present symbol interval is taken to be $t = 0$, it is evident that this pulse will interfere with the pulse in the immediately succeeding symbol interval. Or, in other words, the output at the current sampling instant depends on the present symbol and the preceding symbol.

The transfer function in Eq. (1.3.1) can be expressed as a cascade of a digital filter and an analog ideal LPF, as shown in Fig. 1.4. From Figs. 1.3 and 1.4 we know that at the sampling instants,

$$s_k = q_k + q_{k-1}, \tag{1.3.3}$$

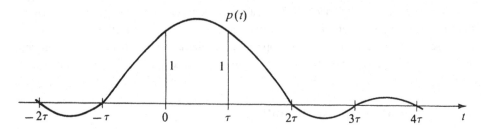

Fig. 1.3 Duobinary impulse response

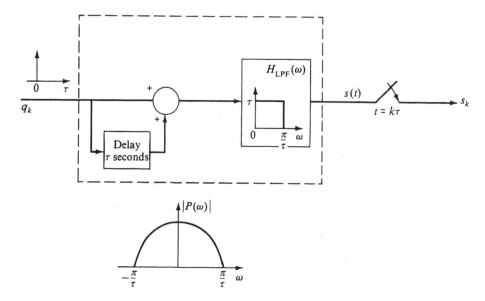

Fig. 1.4 Alternative representation of a duobinary shaping filter

and thus if each member of the $\{q_n\}$ sequence takes on the values $+1$, s_k will be $+2$, 0, or -2. The introduction of dependencies between samples thus changes the binary sequence into a three-level sequence (at the sampling times) for transmission. Decoding is possible, since if q_0 is specified, we can find q_1 using Eq. (1.3.3); with q_1 we can find q_2, and so on. We notice immediately from this decoding scheme that errors will tend to propagate, since if q_{k-1} is erroneously decoded, it is likely that q_k will be decoded incorrectly, and so on. Fortunately, Lender (1963) recognized this problem and corrected it by the use of a technique called *precoding*.

Precoding depends on the particular partial response technique being considered; hence we begin by presenting precoding for duobinary or class 1 partial response schemes. We use precoding to generate an appropriate $\{q_k\}$ sequence to serve as the input to the shaping filter in Fig. 1.4. Therefore, let the original input data sequence be denoted by

$\{a_k\}$, with each a_k taking on the values 0 or 1. We then form a new sequence, denoted $\{b_k\}$, according to

$$b_k = a_k \oplus b_{k-1}, \tag{1.3.4}$$

where the symbol $\oplus$ represents modulo-2 addition. The precoded sequence $\{b_k\}$ still takes on the values 0 or 1, but it is converted into an appropriate sequence q_k by

$$q_k = 2b_k - 1, \tag{1.3.5}$$

using standard base-10 operations. Equation (1.3.3) still represents the duobinary system encoding rule at each sampling instant, so the decoding rule at the receiver is

$$a_k = \begin{cases} 0, & \text{if } s_k = \pm 2 \\ 1, & \text{if } s_k = 0. \end{cases} \tag{1.3.6}$$

Table 1.1 presents an illustrative example of using the precoding operation and of computing the transmitted values at the sampling instants for duobinary signaling. Comparing the $\{a_k\}$ and $\{s_k\}$ sequences in Table 1.1 substantiates the validity of the decoding rule in Eq. (1.3.6). The reader should sketch the superposition of the impulse responses to the $\{q_k\}$ sequence in Table 1.1 to validate the values shown for $s(t)$ at the sampling instants (see Problem 1.14). To illustrate the effect of transmission errors when precoding is used, we assume that the third value of s_k in Table 1.1, s_3, is received as a -2 rather than the correct value 0. It is immediately evident that the only decoding error made is in a_3, which from Eq. (1.3.6) we will erroneously decide was a 0. All other values of a_k are unaffected.

Another practically important correlative coding scheme is class 4 partial response, also called modified duobinary, which has the spectral shaping filter

$$P(\omega) = \begin{cases} j2\tau e^{-j\omega\tau} \sin \omega\tau, & |\omega| \le \frac{\pi}{\tau} \\ 0, & \text{otherwise,} \end{cases} \tag{1.3.7}$$

which can also be rewritten as

Table 1.1 Example of duobinary signaling with precoding

k	0	1	2	3	4	5	6	7	8	9	10	11
a_k		0	0	1	1	0	1	1	1	0	0	1
b_k	1	1	1	0	1	1	0	1	0	0	0	1
q_k	1	1	1	−1	1	1	−1	1	−1	−1	−1	1
s_k		2	2	0	0	2	0	0	0	−2	−2	0

$$P(\omega) = \begin{cases} \tau[1 - e^{-j2\omega\tau}], & |\omega| \leq \frac{\pi}{\tau} \\ 0, & \text{otherwise.} \end{cases} \tag{1.3.8}$$

The class 4 partial response filter impulse response follows immediately from Eq. (1.3.8) as

$$p(t) = \frac{\sin(\pi t/\tau)}{\pi t/\tau} - \frac{\sin[\pi(t - 2\tau)/\tau]}{\pi(t - 2\tau)/\tau}. \tag{1.3.9}$$

For $|\omega| \leq \pi/\tau$, we can factor $P(\omega)$ in Eq. (1.3.8) as $P(\omega) = \tau[1 - e^{-j\omega\tau}] \cdot [1 + e^{-j\omega\tau}]$, which is easily shown to yield

$$P(\omega) = \begin{cases} 2\tau[1 - e^{-j\omega\tau}]e^{-j\omega\tau/2}\cos\frac{\omega\tau}{2}, & |\omega| \leq \frac{\pi}{\tau} \\ 0, & \text{otherwise.} \end{cases} \tag{1.3.10}$$

$P(\omega)$ in Eq. (1.3.10) represents a common implementation of class 4 partial response, which is shown in Fig. 1.5.

If the original data is a sequence of 0's or 1's, denoted $\{a_k\}$, the precoding for class 4 partial response can be expressed as

$$b_k = a_k \oplus b_{k-2}, \tag{1.3.11}$$

and the input to the shaping filter in Fig. 1.5 is thus

$$q_k = 2b_k - 1. \tag{1.3.12}$$

The shaping filter output at the sampling instants is the sequence $\{s_k\}$, where

$$s_k = q_k - q_{k-2}, \tag{1.3.13}$$

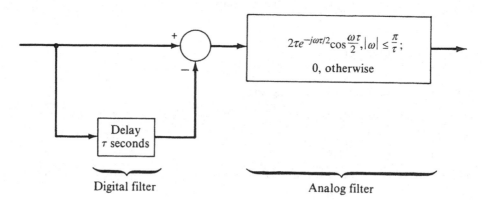

Digital filter   Analog filter

Fig. 1.5 Implementation of class 4 partial response

and the decoding rule in terms of the sample values is

$$a_k = \begin{cases} 0, \text{ if } s_k = 0 \\ 1, \text{ if } s_k = \pm 2. \end{cases} \qquad (1.3.14)$$

Table 1.2 illustrates the use of class 4 partial response signaling to represent the same $\{a_k\}$ sequence encoded by duobinary in Table 1.1. Again, the decoding rule in Eq. (1.3.14) is easily validated by comparing the sequences $\{a_k\}$ and $\{s_k\}$ in Table 1.2.

A block diagram for generalized partial response signaling is shown in Fig. 1.6 (Pasupathy 1977). Precoding for generalized partial response consists of solving the expression

Table 1.2 Example of class 4 partial response signaling with precoding

k	−1	0	1	2	3	4	5	6	7	8	9	10	11
a_k			0	0	1	1	0	1	1	1	0	0	1
b_k	1	1	1	1	0	0	0	1	1	0	1	0	0
q_k	1	1	1	1	−1	−1	−1	1	1	−1	1	−1	−1
s_k			0	0	−2	−2	0	2	2	−2	0	0	−2

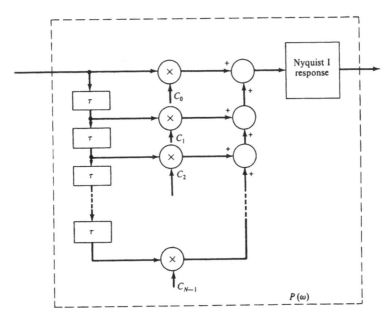

Fig. 1.6 Generalized partial response signaling. From S. Pasupathy, "Correlative Coding: A Bandwidth-Efficient Signaling Scheme", *IEEE Commun. Soc. Mag.*, © 1977 IEEE

$$a_k = \left[C_0 b_k + C_1 b_{k-1} + \cdots + C_{N-1} b_{k-N+1} \right]_{\mathrm{mod}\,2} \tag{1.3.15}$$

for b_k, where $\{a_k\}$ is the input data sequence as before. For duobinary $C_0 = 1$ and $C_1 = 1$, while for class 4 partial response $C_0 = 1$, $C_1 = 0$, and $C_2 = -1$. The reader should use Fig. 1.6 and Eq. (1.3.15) to derive the previously presented equations for duobinary and modified duobinary (see Problem 1.18). Note from Fig. 1.6 that the impulse response of partial response systems is always a weighted superposition of delayed $\sin(\pi t/\tau)/(\pi t/\tau)$ functions. This is clearly evident from Eq. (1.3.9) for class 4 partial response.

To decode, we first note that the transmitted signal $s(t)$ in Fig. 1.6 when evaluated at the sampling times is

$$s_k = C_0 q_k + C_1 q_{k-1} + \cdots + C_{N-1} q_{k-N+1}, \tag{1.3.16}$$

so

$$q_k = \frac{1}{C_0} \left[s_k - \sum_{i=1}^{N-1} C_i q_{k-i} \right]. \tag{1.3.17}$$

Using Eq. (1.3.12), we find that

$$b_k = \frac{1}{2}[q_k + 1]. \tag{1.3.18}$$

Since all preceding b_j's, $j = k - N + 1, \ldots, k - 1$, are known, we can substitute these values along with b_k from Eq. (1.3.18) into Eq. (1.3.15) to obtain a_k.

Correlative coding systems can also be defined for M-level (rather than binary) input data sequences. For this case, the number of transmitted levels is $2M - 1$ and the operations for precoding and decoding are modulo-M. For more information on these methods, see Lender (1981).

Some partial response spectral shaping functions have nulls at dc as well as at $\omega = \pi/\tau$. In particular, as can be seen from Eq. (1.3.7), the class 4 partial response shaping filter has a spectral null at $\omega = 0$. This property can be especially important for data transmission over transformer-coupled telephone lines and when data are to be transmitted using single-sideband modulation.

Since there is correlation among the transmitted levels at sampling instants, this correlation, or memory, can be utilized to detect errors without the insertion of redundant bits at the transmitter. For example, for duobinary signaling the following properties can be discerned: (1) an extreme level (± 2) follows another extreme level of the same polarity only if an even number of zero levels occur in between, (2) an extreme level follows an extreme level of the opposite polarity only if an odd number of zero levels occur in between, and (3) an extreme level of one polarity cannot be followed by an extreme level of the opposite polarity at the next sampling instant. The reader should verify these properties for Table 1.1. Similar properties can also be stated for class 4 partial response signaling. Although all errors cannot be detected, these properties allow error monitoring without additional complexity.

1.4   Deterministic Distortion

Impairments that can affect data transmission over communications channels can be classified as being one of two types: deterministic impairments or random impairments. *Deterministic impairments* include amplitude distortion, delay distortion, nonlinearities, and frequency offset. These impairments are deterministic in that they can be measured for a particular transmission path and they do not change perceptibly over a relatively short period of time. On the other hand, *random impairments* include Gaussian noise, impulsive noise, and phase jitter. Clearly, these random disturbances require a probabilistic description, and hence are treated in a later chapter.

Amplitude distortion and delay distortion can cause intersymbol interference, which decreases the system margin against noise. The magnitude response of a "typical" telephone channel is sketched in Fig. 1.7. Note that what is shown is not "gain versus frequency", but "attenuation versus frequency". It is quite evident that the amplitude response as a function of frequency between 200 and 3200 Hz is far from flat, and hence certainly not distortionless.

A demonstration of the detrimental effects of a nonflat frequency response is obtainable by considering the classic example of a filter with ripples in its amplitude response. In particular, we wish to examine the time-domain response of a system with the transfer function

$$H(\omega) = [1 + 2\varepsilon \cos \omega t_0] e^{-j\omega t_d}, \tag{1.4.1}$$

where $|\varepsilon| \leq \frac{1}{2}$. It should be evident to the reader that $|H(\omega)|$ has ripples of maximum value $\pm 2\varepsilon$ about 1 and that the phase of $H(\omega)$ is linear. We thus have amplitude distortion only. Rewriting Eq. (1.4.1), we obtain

$$
\begin{aligned}
H(\omega) &= \left[1 + \varepsilon \left(e^{j\omega t_0} + e^{-j\omega t_0} \right) \right] e^{-j\omega t_d} \\
&= e^{-j\omega t_d} + \varepsilon e^{j\omega(t_0 - t_d)} + \varepsilon e^{-j\omega(t_0 + t_d)}.
\end{aligned}
\tag{1.4.2}
$$

For a general input pulse shape, say $x(t)$, the time-domain response of the filter in Eq. (1.4.2) is

$$y(t) = x(t - t_d) + \varepsilon x(t - t_d + t_0) + \varepsilon x(t - t_d - t_0). \tag{1.4.3}$$

The ripples in the magnitude response of $H(\omega)$ have thus generated scaled replicas of $x(t)$ that both precede and follow an undistorted (but delayed) version of the input. These replicas of $x(t)$ have been called "echoes" in the literature. Depending on the value of t_0,

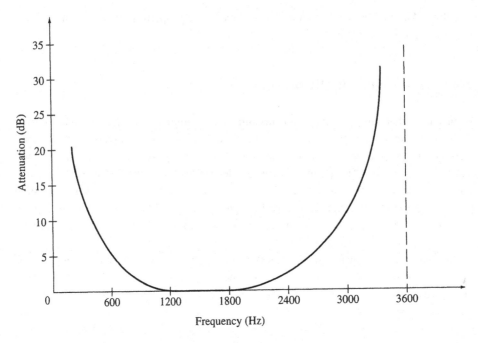

Fig. 1.7 Magnitude response of a typical analog telephone channel

we find that the echoes can interfere with adjacent transmitted pulses, yielding intersymbol interference, and/or the echoes can overlap the main output pulse $x(t - t_d)$, also causing distortion. Sketches of these various cases are left as a problem.

A filter (or channel) with ripples in the phase but a flat magnitude response will also cause echoes. To demonstrate this claim, we consider the transfer function

$$H(\omega) = \exp\{-j[\omega t_d - \varepsilon \sin \omega t_0]\} \qquad (1.4.4)$$

with $|\varepsilon| \ll \pi$. To proceed, we note that $e^{-j\omega t_d}$ represents a pure time delay and that

$$e^{j\varepsilon \sin \omega t_0} = \cos[\varepsilon \sin \omega t_0] + j \sin[\varepsilon \sin \omega t_0]. \qquad (1.4.5)$$

Since $|\varepsilon| \ll \pi$, we can approximate Eq. (1.4.5) as

$$e^{j\varepsilon \sin \omega t_0} \cong 1 + j\varepsilon \sin \omega t_0, \qquad (1.4.6)$$

so

$$H(\omega) \cong e^{-j\omega t_d}\left[1 + \frac{\varepsilon}{2}e^{j\omega t_0} - \frac{\varepsilon}{2}e^{-j\omega t_0}\right]. \qquad (1.4.7)$$

For a general input pulse shape $x(t)$, then, the filter output is approximately

$$y(t) \cong x(t - t_d) + \frac{\varepsilon}{2}x(t - t_d + t_0) - \frac{\varepsilon}{2}x(t - t_d - t_0) \qquad (1.4.8)$$

and again we have echoes.

There is often considerable phase nonlinearity over communications channels, and as we have seen, data transmission can be impaired significantly. Phase information for a communications channel may not be expressed as a system phase response, due to the difficulty of establishing an absolute phase reference and the necessity to count modulo 2π or $360°$. To obtain phase information, a related quantity called *envelope delay* is measured instead.

The envelope delay is defined as the rate of change of the phase versus frequency response, and hence can be expressed as

$$t_R = \frac{-d}{d\omega}\theta(\omega), \qquad (1.4.9)$$

where $\theta(\omega)$ is the channel phase in radians and t_R denotes the envelope delay in seconds. A related quantity is the *phase delay* or *carrier delay*, which is defined as the change in phase versus frequency, and it is expressible as

$$t_c = \frac{-\theta(\omega)}{\omega}, \qquad (1.4.10)$$

where t_c denotes the phase or carrier delay. These definitions are instrumental to the measurement of phase distortion over a channel, and the following derivation illustrates the reasoning behind their names as well as the technique used to measure the envelope delay.

We consider a bandpass channel with a constant magnitude response, $|H(\omega)| = K$, but with a nonlinear phase response, $\angle H(\omega) = \theta(\omega)$. If the input to this channel is a narrowband signal

$$x(t) = m(t)\cos\omega_c t, \qquad (1.4.11)$$

we wish to write an expression for the channel output, denoted $y(t)$, in terms of the envelope delay and the phase delay. To begin, we assume that the phase nonlinearity is not too severe, and thus we can expand the phase in a Taylor's series about the carrier components $\pm\omega_c$. About $\omega = +\omega_c$, we have

$$\theta(\omega) = \theta(\omega_c) + (\omega - \omega_c)\frac{d\theta(\omega)}{d\omega}\bigg|_{\omega=\omega_c}, \qquad (1.4.12)$$

where higher-order terms are assumed negligible. At the frequency of interest, namely $\omega = +\omega_c$, we know that

$$t_R = -\frac{d\theta(\omega)}{d\omega}\bigg|_{\omega=\omega_c} \qquad (1.4.13)$$

and

$$t_c = -\frac{\theta(\omega)}{\omega}\bigg|_{\omega=\omega_c}, \tag{1.4.14}$$

so

$$\theta(\omega) = -\omega_c t_c + (\omega - \omega_c)(-t_R) = -\omega_c t_c - (\omega - \omega_c)t_R. \tag{1.4.15}$$

Similarly, about $\omega = -\omega_c$, we find that

$$\theta(\omega) = \omega_c t_c - (\omega + \omega_c)t_R. \tag{1.4.16}$$

Letting $M(\omega) = \mathcal{F}\{m(t)\}$ and $Y(\omega) = \mathcal{F}\{y(t)\}$, we can write the channel response as

$$\begin{aligned}
Y(\omega) &= \frac{K}{2}[M(\omega - \omega_c) + M(\omega + \omega_c)]e^{j\theta(\omega)} \\
&= \frac{K}{2}\Big[M(\omega - \omega_c)e^{j\theta(\omega)} + M(\omega + \omega_c)e^{j\theta(\omega)}\Big]. \tag{1.4.17}
\end{aligned}$$

Since $M(\omega - \omega_c)$ is located about $\omega = +\omega_c$ and since $M(\omega + \omega_c)$ is located about $\omega = -\omega_c$, we make the appropriate substitutions into Eq. (1.4.17) using Eqs. (1.4.15) and (1.4.16), respectively. Hence

$$\begin{aligned}
Y(\omega) &= \frac{K}{2}\Big[M(\omega - \omega_c)e^{j[-\omega_c t_c - (\omega - \omega_c)t_R]} + M(\omega + \omega_c)e^{j[\omega_c t_c - (\omega + \omega_c)t_R]}\Big] \\
&= \frac{K}{2}e^{-j\omega t_R}\Big[M(\omega - \omega_c)e^{-j\omega_c[t_c - t_R]} + M(\omega + \omega_c)e^{j\omega_c[t_c - t_R]}\Big]. \tag{1.4.18}
\end{aligned}$$

Therefore, using Fourier transform properties we obtain

$$\begin{aligned}
y(t) &= \frac{K}{2}\Big[m(t)e^{j\omega_c[t - t_c + t_R]} + m(t)e^{-j\omega_c\,[t - t_c + t_R]}\Big]\bigg|_{t=t-t_R} \\
&= Km(t - t_R)\cos[\omega_c(t - t_c)]. \tag{1.4.19}
\end{aligned}$$

Only the envelope $m(t)$ is delayed by t_R, hence t_R is called the *envelope delay*, while only the carrier is delayed by t_c, hence the name *carrier delay*.

The method used to measure the envelope delay over real channels satisfies the assumptions used in the derivation. For example, $m(t)$ is selected to be a low-frequency sinusoid, say $\cos \omega_m t$, so that we have $\omega_c \pm \omega_m \cong \omega_c$ and hence the phase nonlinearity will not be too bad over this region. We can also neglect higher-order terms in the Taylor's series since our range of frequencies of interest is $\omega_c - \omega_m \le \omega \le \omega_c + \omega_m$ and ω_m is small. To measure t_R, the derivative is approximated by $\Delta\theta(\omega)/\Delta\omega$, where $\Delta\omega = \omega_m$, and the phase of the received envelope $\cos(\omega_m t - \omega_m t_R)$ is compared to the transmitted envelope to yield $\Delta\theta(\omega) = \omega_m t_R$. Taking the ratio yields the envelope delay.

Results published by Sunde (1961) clearly demonstrate the effects of phase distortion on data transmission. For zero phase distortion, the phase response should be linear, and hence the envelope delay should be constant. Figure 1.8 shows plots of 100% roll-off raised cosine pulses subjected to quadratic envelope delay distortion. To obtain these plots, Sunde (1961) inserted delay distortion, which increased quadratically from 0 at $\omega = 0$ to some final value at $\omega = \omega_{\text{max}}$. As the envelope delay increases, the pulse amplitude is reduced and the pulse peak no longer occurs at the desired sampling instant. Furthermore, the zero crossings become shifted and the trailing pulse becomes large enough in amplitude to interfere with adjacent pulses. Obviously, delay distortion or phase distortion can be exceedingly detrimental in data transmission.

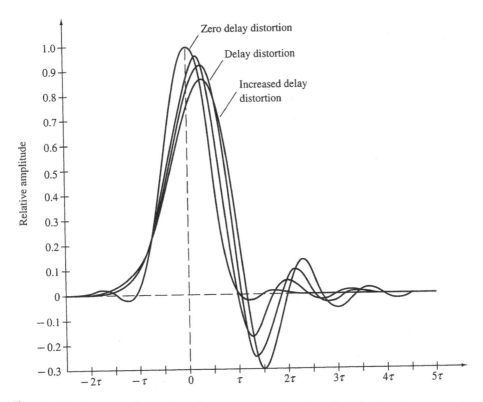

Fig. 1.8 Raised cosine pulses with quadratic delay distortion. From E. D. Sunde, "Pulse Transmission by AM, FM, and PM in the Presence of Phase Distortion", *Bell Syst. Tech. J.*, © 1961 AT&T Bell Laboratories

1.5   Eye Patterns

A convenient way to see the distortion present on a channel is to display what is called the system *eye pattern* or *eye diagram*. The eye pattern is obtained by displaying the data pulse stream on an oscilloscope, with the pulse stream applied to the vertical input and the sampling clock applied to the external trigger. A drawing of a two-level eye pattern is shown in Fig. 1.9, and the source of its name is clearly evident. Typically, one to three pulse (symbol) intervals are displayed and several kinds of distortion are easily observed. For minimum error probability, sampling should occur at the point where the eye is open widest. If all of the traces go through allowable (transmitted) pulse amplitudes only at the sampling instants, the eye is said to be *100% open* or *fully open*. A fully open eye pattern for three-level pulse transmission is sketched in Fig. 1.10. The eye pattern is said to be 80% open if, at the sampling instant, the traces deviate by 20% from the fully open eye diagram. This degradation is sometimes expressed as a loss in signal-to-noise ratio (S/N) by S/N = $20 \log_{10} 0.8 = -1.9$ dB, where the minus sign indicates a degradation in S/N.

The distance between the decision thresholds and adjacent received pulse traces at the sampling time is the margin of the system against additional noise. As the sampling time

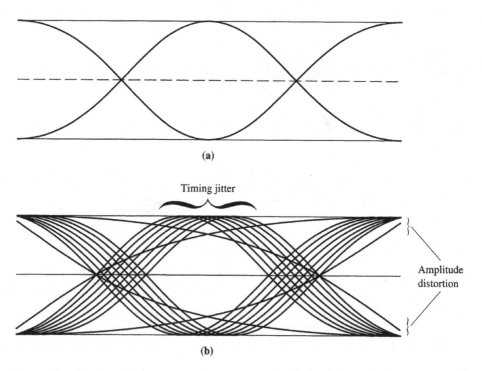

(a)

(b)

Fig. 1.9 Two-level eye diagrams: **a** two-level pattern (distortionless); **b** two-level eye pattern with timing jitter

Fig. 1.10 Three-level eye pattern

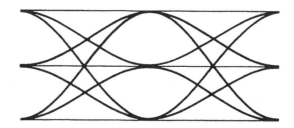

is varied about the time instant of maximum eye opening, the eye begins to close. The rate that the eye closes as the sampling instant is varied is an indication of the system's sensitivity to timing error. Jitter in received zero crossings (or threshold crossings) can be particularly insidious, since many receivers extract timing information by averaging zero crossings. The various kinds of distortion are labeled in Fig. 1.9.

As can be seen from Fig. 1.10, as the number of transmitted levels per pulse is increased for a fixed maximum pulse amplitude, the smaller is the fully open eye's margin against noise. Thus, increasing the number of transmitted levels per pulse, while increasing the transmitted bit rate, also tends to increase the number of bit errors. When the distortion is so severe that the eye is completely filled by pulse traces, the eye is said to be *closed*. Obviously, when the eye is closed, data transmission is extremely unreliable.

1.6   Equalization

For data transmission to be effective, the data must be received accurately. One way to reduce the number of data transmission errors is to correct, or compensate for, the deterministic distortion contributed by the channel. This is easily accomplished if the channel transfer function is known, since if the channel transfer function is $|H_c(\omega)|e^{j\angle H_c(\omega)}$, a cascaded equalizer with transfer function $H_{eq}(\omega) = |H_c(\omega)|^{-1}e^{-j\angle H_c(\omega)}$ will produce an ideal response. This approach is effective for deterministic distortion, since deterministic impairments are assumed to vary slowly in comparison to a single pulse interval. If the channel transfer function is known and the channel response does not vary significantly with time, the equalizer can be preset and held fixed while the data are transmitted. This situation is called *fixed equalization*.

Of course, the channel transfer function may not be known in advance, in which case the channel characteristics must be measured or "learned". Furthermore, the channel response may be changing slowly but continually with time. If the channel transfer function is unknown, but it can be assumed to be invariant over a relatively long time period, the channel transfer function can be learned during an initialization or startup period by transmitting a previously agreed upon data sequence known to both the transmitter and receiver, and then held fixed during actual data transmission. Equalizers that use this approach are called *automatic equalizers*. If the channel is unknown and slowly

time varying, adaptation or learning must continue even after the startup sequence, while
the data are being transmitted. Equalizers based on this method are called *adaptive equalizers*. The algorithms used for automatic and adaptive equalization are the same except
for the reference signals employed.

A block diagram of a communication system with equalization is shown in Fig. 1.11.
Modulation and demodulation blocks are not included in this figure and our initial development emphasizes baseband equalization. Passband equalization is discussed at the end
of this section. Throughout our development the channel is assumed to be bandlimited to
B Hz. Equalizers in data transmission applications typically take the form of a transversal filter. A $2N + 1$ tap transversal filter is illustrated in Fig. 1.12. The coefficients
$\{c_i, i = -N, -N + 1, \ldots, 0, 1, \ldots, N - 1, N\}$ are called *equalizer tap gains*, and the
time delays (Δ) are the *tap spacing*. The tap spacing is often fixed at one symbol interval,
while the tap gains are adjustable. The equalizer impulse response is

$$h_{eq}(t) = \sum_{n=-N}^{N} c_n \delta(t - n\Delta), \qquad (1.6.1)$$

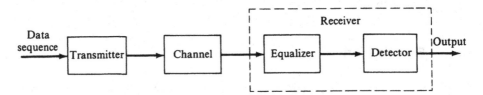

Fig. 1.11 Data communication system with equalization

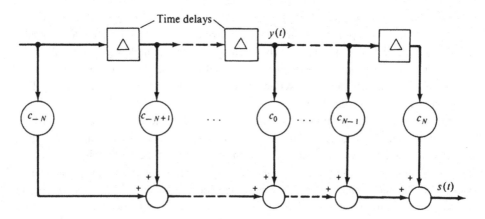

Fig. 1.12 A $2N + 1$ tap equalizer

which has the transfer function

$$H_{eq}(\omega) = \sum_{n=-N}^{N} c_n e^{-j\omega n \Delta}. \tag{1.6.2}$$

To illustrate how a transversal filter structure can be used for channel equalization, consider the following example.

Example 1.6.1 Consider a data transmission channel with the known transfer function

$$H_c(\omega) = \left[a_1 + a_2 e^{-j\omega(t_2 - t_1)} \right] e^{-j\omega t_1}$$

with $t_2 > t_1$ and $(a_2/a_1)^2 \ll 1$. We wish to find $H_{eq}(\omega) = 1/H_c(\omega)$, and then determine c_{-1}, c_0, c_1, and Δ for a three-tap transversal filter equalizer that approximates $H_{eq}(\omega)$. Straightforwardly, we have

$$H_{eq}(\omega) = \frac{1}{H_c(\omega)} = \frac{e^{j\omega t_1}}{a_1} \cdot \frac{1}{1 + a_2 e^{-j\omega(t_2-t_1)}/a_1}. \tag{1.6.3}$$

Since $(a_2/a_1)^2 \ll 1$, $a_2/a_1 < 1$, and we can use a series expansion to approximate $H_{eq}(\omega)$ as

$$H_{eq}(\omega) \cong \frac{e^{j\omega t_1}}{a_1} \left[1 - \frac{a_2}{a_1} e^{-j\omega(t_2-t_1)} + \left(\frac{a_2}{a_1} \right)^2 e^{-j2\omega(t_2-t_1)} \right]. \tag{1.6.4}$$

Letting $N = 1$ in Eq. (1.6.2), we know that for a transversal filter equalizer

$$H_{eq}(\omega) = c_{-1} e^{j\omega \Delta} + c_0 + c_1 e^{-j\omega \Delta} = e^{j\omega \Delta} \left[c_{-1} + c_0 e^{-j\omega \Delta} + c_1 e^{-j2\omega \Delta} \right]. \tag{1.6.5}$$

Comparing Eqs. (1.6.4) and (1.6.5), and ignoring the terms in front of the brackets, which involve only gain elements and time shifts, we find that

$$\Delta = t_2 - t_1, \quad c_{-1} = 1, \quad c_0 = \frac{-a_2}{a_1}, \quad c_1 = \left(\frac{a_2}{a_1} \right)^2.$$

Thus a three-tap transversal filter with these parameter values will approximately equalize $H_c(\omega)$.

Let us now proceed to investigate the transversal filter as an automatic equalizer. For $y(t)$ in, as shown in Fig. 1.12, the equalizer output is

$$s(t) = y(t) * \sum_{n=-N}^{N} c_n \delta(t - n\Delta) = \sum_{n=-N}^{N} c_n y(t - n\Delta). \tag{1.6.6}$$

If we consider only the sampling times $t = k\Delta$, then Eq. (1.6.6) becomes

$$s(k\Delta) = \sum_{n=-N}^{N} c_n y[(k-n)\Delta].$$ (1.6.7)

It is common notational practice to drop the explicit dependence on Δ and to write

$$s_k = \sum_{n=-N}^{N} c_n y_{k-n} = c_{-N} y_{k+N} + \cdots + c_{-1} y_{k+1} + c_0 y_k + c_1 y_{k-1} + \cdots + c_N y_{k-N},$$

(1.6.8)

where the dependence on Δ is implicit. For a single positive unit-amplitude raised cosine pulse transmitted at time instant (sampling instant) k, the desired sample values at the equalizer output will be $s_k = 1, s_{k+j} = 0$ for $j \neq 0$. Thus, if the equalizer input $y(t)$ is received undistorted, we have $y_k = 1, y_{k+j} = 0$ for $j \neq 0$, and it is evident that the desired output pulse is centered in time on the equalizer taps.

Although not indicated in Fig. 1.12, the equalizer tap gains are adjustable and must be calculated from measurements taken over the channel in question. This is normally accomplished by transmitting a fairly long sequence of pulses, but for purposes of illustration, let us consider the case of a single raised cosine pulse transmitted at time instant $k = 0$. Any other time instant could be used, but $k = 0$ simplifies notation. The following example illustrates how the equalizer tap gains can be calculated from the transmission of this single pulse over the channel and a known reference signal.

Example 1.6.2 Consider a three-tap equalizer ($N = 1$) with the input sequence $y_1 = -1/2$, $y_0 = 1, y_{-1} = 1/4$ and all other $y_k = 0$. Using Eq. (1.6.8), we can find the equalizer output for all values of k as:

$$k \leq -3 : s_k = c_{-1} y_{k+1} + c_0 y_k + c_1 y_{k-1} = 0$$
$$k = -2 : s_{-2} = c_{-1} y_{-1} + c_0 y_{-2} + c_1 y_{-3} = c_{-1} y_{-1} = \tfrac{1}{4} c_{-1}$$
$$k = -1 : s_{-1} = c_{-1} y_0 + c_0 y_{-1} + c_1 y_{-2} = c_{-1} y_0 + c_0 y_{-1} = c_{-1} + \tfrac{1}{4} c_0$$
$$k = 0 \ \ : s_0 = c_{-1} y_1 + c_0 y_0 + c_1 y_{-1} = -\tfrac{1}{2} c_{-1} + c_0 + \tfrac{1}{4} c_1$$
$$k = 1 \ \ : s_1 = c_{-1} y_2 + c_0 y_1 + c_1 y_0 = -\tfrac{1}{2} c_0 + c_1$$
$$k = 2 \ \ : s_2 = c_{-1} y_3 + c_0 y_2 + c_1 y_1 = -\tfrac{1}{2} c_1$$
$$k \geq 3 \ \ : s_k = 0.$$

If the input to the channel is a single raised cosine pulse at time instant $k = 0$, the desired output of the equalizer is $r_0 = 1, r_k = 0$ for all other k. The equations to be solved for the tap gains are thus

$$\frac{1}{4} c_{-1} = 0$$ (1.6.9)

$$c_{-1} + \frac{1}{4}c_0 = 0 \qquad (1.6.10)$$

$$-\frac{1}{2}c_{-1} + c_0 + \frac{1}{4}c_1 = 1 \qquad (1.6.11)$$

$$-\frac{1}{2}c_0 + c_1 = 0 \qquad (1.6.12)$$

$$-\frac{1}{2}c_1 = 0. \qquad (1.6.13)$$

We see immediately that these equations are inconsistent.

Since we have only three equalizer coefficients at our disposal, it is unrealistic to expect that we can fix the output at five sampling instants. Hence, to determine the tap gains we retain only the three equations centered on the time instant $k = 0$, Eqs. (1.6.10)–(1.6.12). Solving these equations simultaneously, we find that $c_{-1} = -\frac{1}{5}$, $c_0 = \frac{4}{5}$, and $c_1 = \frac{2}{5}$. For the given equalizer input sequence $\{y_k\}$, we can use Eq. (1.6.8) and the computed tap gains to show that the equalizer output for all k is $s_{k-2} = -\frac{1}{20}$, $s_{k-1} = 0$, $s_k = 1$, $s_{k+1} = 0$, $s_{k+2} = -\frac{1}{5}$, and $s_k = 0$ for all other k. Thus the equalizer has produced an ideal response at time instants $k - 1$, k, and $k + 1$, but the outputs at times $k - 2$ and $k + 2$ are still nonzero, since they cannot be controlled by the equalizer. Since the outputs at the time instants other than k are zeroed, this type of equalizer is called a *zero-forcing* (ZF) equalizer.

Before proceeding with the discussion of equalization, we note that the expression for the equalizer output in Eq. (1.6.8) is a discrete-time convolution, which can also be expressed as a polynomial multiplication. To do this, we write the input sequence $\{y_k\}$ as a polynomial by associating the variable p^j with y_j. Similarly, we associate the variable p^k with c_k to obtain a polynomial from the tap gains. To demonstrate the procedure, we consider the equalizer input sequence from Example 1.6.2, namely $y_1 = -\frac{1}{2}$, $y_0 = 1$, $y_{-1} = \frac{1}{4}$, $y_j = 0$ for all other j, and the resulting computed tap gains $c_{-1} = -\frac{1}{5}$, $c_0 = \frac{4}{5}$, and $c_1 = \frac{2}{5}$. The polynomial multiplication thus becomes

$$\left(y_{-1}p^{-1} + y_0 p^0 + y_1 p^1\right)\left(c_{-1}p^{-1} + c_0 p^0 + c_1 p^1\right)$$
$$= c_{-1}y_{-1}p^{-2} + (c_0 y_{-1} + y_0 c_{-1})p^{-1}$$
$$+ (y_{-1}c_1 + c_0 y_0 + c_{-1}y_1)p^0 + (y_0 c_1 + c_0 y_1)p^1 + y_1 c_1 p^2. \qquad (1.6.14)$$

The output at time s_{k+j} is the coefficient of p^j; therefore, upon substituting for the equalizer input and tap gains, we find that $s_{k-2} = -\frac{1}{20}$, $s_{k-1} = 0$, $s_k = 1$, $s_{k+1} = 0$, $s_{k+2} = -\frac{1}{5}$, with all other $s_n = 0$, since these terms are absent from Eq. (1.6.14). Of course, these results are the same as those we obtained using Eq. (1.6.8).

Another type of equalization does not attempt to force the equalizer outputs to zero at all time instants (within its range) except k, but instead, tries to minimize the sum of the squared errors over a set of output samples. Specifically, in minimum mean-squared error (MMSE) equalization, the tap gains are selected to minimize

$$
\varepsilon^2 = \sum_{k=-K}^{K} e_k^2 = \sum_{k=-K}^{K} \left[r_k - \sum_{n=-N}^{N} c_n y_{k-n} \right]^2, \tag{1.6.15}
$$

where $\{r_k\}$ is the desired equalizer output or reference sequence, c_n, N, and $\{y_k\}$ are as defined previously, and K is to be selected. Taking partial derivatives with respect to the c_m, we have

$$
\frac{\partial \varepsilon^2}{\partial c_m} = 0 = -2 \sum_{k=-K}^{K} \left[r_k - \sum_{n=-N}^{N} c_n y_{k-n} \right] y_{k-m} \tag{1.6.16}
$$

for $m = -N, -N+1, \ldots, -1, 0, 1, \ldots, N-1, N$ or equivalently,

$$
\sum_{n=-N}^{N} c_n \sum_{k=-K}^{K} y_{k-n} y_{k-m} = \sum_{k=-K}^{K} r_k y_{k-m}. \tag{1.6.17}
$$

Since the sequences $\{y_j\}$ and $\{r_j\}$ are known, we can compute $\sum_{k=-K}^{K} y_{k-n} y_{k-m}$ and $\sum_{k=-K}^{K} r_k y_{k-m}$ for all m and n, so that from Eq. (1.6.17) we have a set of $2N+1$ simultaneous equations in $2N+1$ unknowns to be solved for the tap gains. To illustrate this procedure, we compute the MMSE equalizer coefficients for the same equalizer input sequence used in Example 1.6.2.

Example 1.6.3 The equalizer input sequence is $y_{-1} = \frac{1}{4}$, $y_0 = 1$, $y_1 = -\frac{1}{2}$, with all other $y_j = 0$, and the reference sequence is $r_0 = 1$, $r_j = 0$ for all $j \neq 0$. For $N = 1$ we have to solve the three simultaneous equations represented by [from Eq. (1.6.17)],

$$
\sum_{n=-1}^{1} c_n \sum_{k=-K}^{K} y_{k-n} y_{k-m} = \sum_{k=-K}^{K} r_k y_{k-m} \tag{1.6.18}
$$

with $m = -1, 0, 1$, for the tap gains $\{c_n\}$. It is instructive to write these equations out in detail. The three simultaneous equations represented by Eq. (1.6.18) are

$m = -1$:

$$
c_{-1} \overset{①}{\sum_k y_{k+1}^2} + c_0 \overset{②}{\sum_k y_k y_{k+1}} + c_1 \overset{③}{\sum_k y_{k-1} y_{k+1}} = \sum_k r_k y_{k+1} \tag{1.6.19}
$$

$m = 0$:

$$c_{-1} \overset{④}{\sum_k y_{k+1} y_k} + c_0 \overset{⑤}{\sum_k y_k^2} + c_1 \overset{⑥}{\sum_k y_{k-1} y_k} = \sum_k r_k y_k \qquad (1.6.20)$$

$m = 1$:

$$c_{-1} \overset{⑦}{\sum_k y_{k+1} y_{k-1}} + c_0 \overset{⑧}{\sum_k y_k y_{k-1}} + c_1 \overset{⑨}{\sum_k y_{k-1}^2} = \sum_k r_k y_{k-1}, \qquad (1.6.21)$$

where the summations on k are from $-K$ to K. Normally, the parameter K has to be prese-lected; however, for our simple example we need only assume that K is large. More will be said on this point later.

To solve Eqs. (1.6.19)—(1.6.21) for the tap gains, we first need to evaluate the terms involving the summations on k. Since $r_j = 0$ for $j \neq 0$, the summations on the right side of the equals sign in these equations are easily evaluated as $\sum_k r_k y_{k+1} = y_1$, $\sum_k r_k y_k = y_0$, and $\sum_k r_k y_{k-1} = y_{-1}$, for $m = -1, 0$, and 1, respectively. For K large it is also easy to see that we do not have to evaluate all nine summations for the terms on the left sides of the equals signs in Eqs. (1.6.19)—(1.6.21), since many of the summations are the same. Specifically, we can show that ① = ⑤ = ⑨, ② = ④ = ⑥ = ⑧ and ③ = ⑦. Thus we only need to calculate three summations, which are given by

$$① = \sum_k y_{k+1}^2 = y_{-1}^2 + y_0^2 + y_1^2 = 1\frac{5}{16} = ⑤ = ⑨$$

$$② = \sum_k y_k y_{k+1} = y_{-1} y_0 + y_0 y_1 = -\frac{1}{4} = ④ = ⑥ = ⑧$$

$$③ = \sum_k y_{k-1} y_{k+1} = y_{-1} y_1 = -\frac{1}{8} = ⑦.$$

Substituting for all of the values of the summations into Eqs. (1.6.19)–(1.6.21), we must solve the three equations

$$\frac{21}{16} c_{-1} - \frac{1}{4} c_0 - \frac{1}{8} c_1 = -\frac{1}{2} \qquad (1.6.22)$$

$$-\frac{1}{4} c_{-1} + \frac{21}{16} c_0 - \frac{1}{4} c_1 = 1 \qquad (1.6.23)$$

$$-\frac{1}{8} c_{-1} - \frac{1}{4} c_0 + \frac{21}{16} c_1 = \frac{1}{4} \qquad (1.6.24)$$

for c_{-1}, c_0, and c_1. We find that $c_{-1} = -0.2009$, $c_0 = +0.7885$, and $c_1 = 0.3215$.

By substituting for $\{r_k\}$, $\{y_k\}$, and $\{c_k\}$ in Eq. (1.6.15), we can calculate the minimum mean-squared error to be

$$\varepsilon_{\min}^2 = \left[c_{-1}y_{-1}\right]^2 + \left[c_{-1}y_0 + c_0y_{-1}\right]^2 + \left[r_0 - c_{-1}y_1 - c_0y_0 - c_1y_{-1}\right]^2$$
$$+ \left[c_0y_1 + c_1y_0\right]^2 + \left[c_1y_1\right]^2 = 0.0352. \tag{1.6.25}$$

Using the computed tap gains, we can also find the equalizer output sequence to be $s_k = 0$ for $k \leq -3$ and $k \geq 3$, $s_{-2} = -0.0502$, $s_{-1} = -0.0038$, $s_0 = 0.9694$, $s_1 = -0.0727$, $s_2 = -0.1608$. Clearly, the MMSE equalizer has not forced any of the output values to zero; however, no other three-tap equalizer can produce a smaller $\varepsilon_{\min}^2$. It should also be noted that output values outside the equalizer's span are not necessarily small, as is seen from the value of s_2.

We have introduced the concepts of ZF equalizers and MMSE equalizers, and we have shown that equalization consists primarily of solving a set of linear simultaneous equations. For automatic and adaptive equalization in data transmission systems, it has proven more advantageous to solve the necessary set of linear simultaneous equations using iterative methods. Generally, iterative methods consist of making incremental adjustments in the equalizer tap gains after each new pulse is received. These iterative methods are best motivated by reconsidering MMSE equalization and Eqs. (1.6.15) and (1.6.16).

The partial derivative in Eq. (1.6.16) is called the gradient of ε^2, and it obviously represents the rate of change of ε^2 with respect to the tap gains $\{c_m\}$. For a minimum of ε^2, we will have a zero rate of change or zero slope, and therefore we equate this rate of change to zero as in Eq. (1.6.16). If, however, we wish to equalize the channel over a long sequence of pulses and we want to do this one sample at a time, rather than wait until after the entire sequence is transmitted, we consider e_k^2 instead of ε^2. From Eq. (1.6.15) we have that

$$e_k^2 = \left[r_k - \sum_{n=-N}^{N} c_n y_{k-n}\right]^2. \tag{1.6.26}$$

We desire to adjust each of the $2N + 1$ equalizer coefficients at each time instant k to minimize e_k^2 in Eq. (1.6.26). To accomplish this goal, we adjust the tap gains by a small amount in a direction opposite to the slope or gradient of e_k^2 with respect to the $\{c_m\}$. Therefore, we are always adjusting the coefficients toward the minimum of e_k^2.

Taking the gradient (partial derivative) of e_k^2 with respect to each of the $\{c_m\}$, we are thus led to construct the iterative MMSE equalizer adjustment algorithm

$$c_m(k) = c_m(k-1) + G\{r_{k-1} - s_{k-1}\}y_{k-m-1}$$
$$= c_m(k-1) + G\left\{r_{k-1} - \sum_{n=-N}^{N} c_n(k-1)y_{k-n-1}\right\}y_{k-m-1} \tag{1.6.27}$$

for $m = -N, \ldots, -1, 0, 1, \ldots, N$, where the time variation of the tap gains is indicated parenthetically and G is a gain constant to be selected. By incrementally changing the tap gains at each time instant k according to Eq. (1.6.27), we minimize e_k^2 in Eq. (1.6.26), and hence approximately solve the set of linear simultaneous equations represented by Eq. (1.6.17). A similar iterative approach can be used for zero-forcing equalization.

As presently written, the equalizer algorithm in Eq. (1.6.27) is applicable to automatic equalization during startup, but not adaptive equalization during data transmission. This is because of the presence of the known reference sequence $\{r_k\}$. Obviously, during actual data transmission, the desired equalizer output is not known. As a result, for adaptive equalization, $\{r_k\}$ is replaced by an estimated reference, say $\{\hat{r}_k\}$, which is derived from the equalizer output sequence $\{s_k\}$. This $\{\hat{r}_k\}$ is always available and equalization can proceed. Since $\hat{r}_k$ is computed from s_k, the equalizer adjustment is called *decision directed*. The reasoning behind using a version of the past equalizer outputs as a reference is that immediately after startup, the equalizer should be well adjusted and the distortion should be small. If the channel response is changing slowly with time, the derived reference should be accurate, and the equalizer can track these slow changes.

A further extension to the decision-directed concept is the use of decision feedback equalization (DFE). In DFE, not only are previously decided symbols used as reference symbols in the adaptive algorithms but they are fed back through a transversal filter and the resulting feedback transversal filter output is subtracted from the equalizer output. The idea here is that once output symbols are detected, their contribution to the equalizer output can be removed. Decision feedback equalizers are nonlinear filters and their development is left to the literature.

When a modulation method such as QAM is used, the equalizer may be implemented in the passband rather than at baseband, as has been discussed here. An advantage of passband equalization is that rapid phase changes can be tracked more easily, since equalizer delay is reduced.

For all of the equalizers discussed in this section, the tap spacing Δ is chosen to be one symbol interval. It is also possible to design fractionally spaced equalizers with tap spacing equal to some fraction of the symbol interval (usually, one-half). Such fractionally spaced equalizers are less sensitive to sampler phase, can accommodate more delay distortion, and can compensate for amplitude distortion with less noise than symbol-spaced equalizers.

The choice among the various equalizer algorithms depends on the requirements of the specific application, particularly desired convergence rate, initial distortion, desired minimum distortion, and complexity. There is a vast literature on equalization, but the original papers by Lucky (1965, 1966) and Lucky and Rudin (1967) and the book by Lucky et al. (1968) are still useful. Hirsch and Wolf (1970) present a readable performance comparison of some adaptive equalization algorithms, including the ZF and MMSE algorithms. Pahlavan and Holsinger (1988) provide a historical overview of modem evolution, including equalization, and Qureshi (1985) gives an excellent development of adaptive equalization and digital receiver structures.

1.7   Data Scramblers

The performance of data transmission systems must be independent of the specific bit sequence being transmitted. If allowed to occur, repeated bit sequences can cause wide variations in the received power level as well as difficulties for adaptive equalization and clock recovery. Since all these problems are eliminated if the bit sequence is "random" (has no discernible pattern), many modems employ a *data scrambler* to produce a pseudo-random sequence for any given input bit sequence. The scrambler usually takes the form of a shift register with feedback connections, while the unscrambler is a feedforward-connected shift register. The following example clearly illustrates the operation of data scramblers.

Example 1.7.1 A data scrambler and unscrambler are shown in Fig. 1.13. The scrambler operates in the following fashion. The initial shift register contents are arbitrary but prespecified and fixed to be the same in both the scrambler and unscrambler. The first bit in sequence s_1 (note that the subscript does not indicate a time sequence here but simply denotes bit sequence number 1) is summed modulo-2 with the modulo-2 sum of locations 2 and 5 in the shift register. This sum becomes the first bit in bit sequence s_2. As this bit is presented to the channel, the contents of the shift register are shifted up one stage as follows: $5 \rightarrow$ out, $4 \rightarrow 5$, $3 \rightarrow 4$, $2 \rightarrow 3$, $1 \rightarrow 2$. The first bit in s_2 is also placed in shift register stage 1. The next bit of sequence s_1 arrives, and the procedure is repeated.

The unscrambler operates as follows. The initial contents of the shift register are fixed. When the first bit of sequence s'_2 arrives, this bit is summed mod-2 with the mod-2 sum of the initial values of stages 2 and 5. This sum then becomes the first bit of sequence

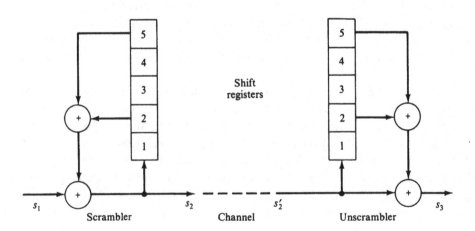

Fig. 1.13 Data scrambler and unscrambler for Example 1.7.1

Table 1.3 Scrambled bit sequence for Example 1.7.1

Time	1	2	3	4	5	6	7	8	9	10	11	12
s_1	1	0	0	0	0	0	0	0	0	0	0	0
1	0	1	0	1	0	1	1	1	0	1	1	0
2	0	0	1	0	1	0	1	1	1	0	1	1
3	0	0	0	1	0	1	0	1	1	1	0	1
4	0	0	0	0	1	0	1	0	1	1	1	0
5	0	0	0	0	0	1	0	1	0	1	1	1
s_2	1	0	1	0	1	1	1	0	1	1	0	0

s_3. At this time instant the contents of the shift register are shifted up one as follows: $5 \rightarrow$ out, $4 \rightarrow 5$, $3 \rightarrow 4$, $2 \rightarrow 3$, $1 \rightarrow 2$. The first bit of sequence s_2' is then put in stage 1, and the next bit in s_2' is presented to the unscrambler. The procedure is then repeated.

As an illustration, consider the input sequence s_1 and the initial shift register contents shown in Table 1.3 for the scrambler of Fig. 1.13. Following the procedure previously outlined for the scrambler, we can compute the scrambler output sequence and shift register contents as shown in the table. Note that in Table 1.3, the output bit at any time instant is computed from the current input bit and the current shift register contents. A shift then takes place and the shift register contents at the next time instant are generated. By inspection of s_2, we can see clearly that s_2 is very different from s_1. Whether s_2 has any special randomness properties is not immediately evident.

Assuming that no bit errors occur on the channel, we have $s_2' = s_2$, and Table 1.4 illustrates the operation of the unscrambler. We compute s_3 from the current input bit (s_2') and the current shift register contents. A shift is then performed to prepare the shift register for the next bit in s_2'. Note that since $s_1 = s_3$, the data are unscrambled.

We have not said anything as yet about how the scrambler and unscrambler circuits are selected. To begin this discussion, we introduce the unit delay operator D, which

Table 1.4 Unscrambled bit sequence for Example 1.7.1

Time	1	2	3	4	5	6	7	8	9	10	11	12
s_2'	1	0	1	0	1	1	1	0	1	1	0	0
1	0	1	0	1	0	1	1	1	0	1	1	0
2	0	0	1	0	1	0	1	1	1	0	1	1
3	0	0	0	1	0	1	0	1	1	1	0	1
4	0	0	0	0	1	0	1	0	1	1	1	0
5	0	0	0	0	0	1	0	1	0	1	1	1
s_3	1	0	0	0	0	0	0	0	0	0	0	0

represents delaying the sequence by one bit. Thus, in this notation, Ds_2 represents the contents of stage 1 in the scrambler shift register, $D^2 s_2$ represents stage 2, and so on. We can thus represent the scrambler circuit in Fig. 1.13 as

$$s_2 = D^2 s_2 \oplus D^5 s_2 \oplus s_1, \tag{1.7.1}$$

where the terms in D, D^3, and D^4 are absent due to the lack of feedback connections at these stages. Taking all terms in s_2 to the left side of the equality, we obtain

$$\left[1 \oplus D^2 \oplus D^5\right] s_2 = s_1, \tag{1.7.2}$$

or in the form of a transfer function relationship,

$$s_2 = \frac{1}{1 \oplus D^2 \oplus D^5} s_1. \tag{1.7.3}$$

By inspection of Eq. (1.7.3), we do not get an indication that a single bit produces a much longer sequence, as is evident in Table 1.3. To see this effect, we can perform synthetic division on Eq. (1.7.3) to reveal

$$s_2 = \left[1 \oplus D^2 \oplus D^4 \oplus D^5 \oplus D^6 \oplus D^8 \oplus D^9 \oplus \cdots\right] s_1. \tag{1.7.4}$$

Hence, from Eq. (1.7.4), we can generate the s_2 sequence from s_1 without the shift register stages.

A representation for the unscrambler can be written similarly as

$$s_3 = \left[1 \oplus D^2 \oplus D^5\right] s_2'. \tag{1.7.5}$$

Note that for no channel errors, $s_2' = s_2$, and if we substitute s_2 in Eq. (1.7.3) into Eq. (1.7.5), we find that $s_3 = s_1$.

This is a special case of a more general result that if we have representations for two circuits as

$$s_2 = F(D)s_1 \tag{1.7.6}$$

and

$$s_3 = G(D)s_2, \tag{1.7.7}$$

these two circuits can be used as a scrambler/unscrambler pair whenever

$$F(D)G(D) = 1. \tag{1.7.8}$$

Thus any pair of feedback- and feedforward-connected shift registers that satisfy Eq. (1.7.8) are suitable for use as a scrambler and an unscrambler pair. In Fig. 1.13 we chose the feedback connection as the scrambler and the feedforward device as the unscrambler. Equation (1.7.8) indicates that we could have used the feedforward connection for scrambling and the feedback connection for unscrambling.

A primary reason for the choice in Fig. 1.13 is bit error propagation. A single bit error into a feedforward connection affects a successive number of bits equal to the shift register length, while for a feedback connection, the effect can be much longer.

For any given shift register length M, there are obviously 2^M possible linear mod-2 sums that can be formed from its contents. All of these connections will not produce a "good" pseudorandom sequence. Furthermore, the output of any linear shift register connection with M stages is periodic with a period of $2^M - 1$ or less. An output sequence with period $2^M - 1$ is a special sequence and is called a *maximal-length linear shift register sequence* (Golomb 1964). For the linear feedback shift register connection serving as a scrambler in Fig. 1.13, a maximal-length sequence would have a period of $2^5 - 1 = 31$.

Further discussion of scramblers, unscramblers, error propagation, and maximal-length sequences is deferred to the problems and Appendix G.

1.8 Carrier Acquisition and Symbol Synchronization

Carrier acquisition and symbol synchronization or timing extraction are critical to the operation of high-performance modems. In this section we discuss briefly some of the many possible approaches for obtaining this information.

For noncoherent modems, such as those employing FSK, carrier acquisition is unnecessary. Further, we have seen that the very popular DPSK systems circumvent coherent reference difficulties by using the received carrier from the immediately preceding symbol interval. Important modulation methods such as VSB and QAM do require a coherent reference to be acquired in some fashion.

Early modems that employed VSB modulation transmitted a carrier tone in phase quadrature or transmitted pilot tones at the edges of the data spectrum. For the latter method, the received signal can be represented as (Lucky et al. 1968)

$$s(t) = m(t) \cos[\omega_c t + \Delta\omega t + \theta(t) + \phi] + \hat{m}(t) \sin[\omega_c t + \Delta\omega t + \theta(t) + \phi]$$
$$+ \alpha \cos[\omega_L t + \Delta\omega t + \theta(t)] + \alpha \cos[\omega_H t + \Delta\omega t + \theta(t)], \qquad (1.8.1)$$

where $\Delta\omega$ denotes frequency distortion, $\theta(t)$ denotes phase jitter, and ω_L and ω_H denote the lower and upper pilot tones. The pilot tone components in Eq. (1.8.1) are removed by narrowband filters, multiplied together, and low-pass filtered to yield a cosine term of frequency $\omega_H - \omega_L$. This term is then fed to a phase-locked loop. During the startup time a carrier tone is transmitted and fixed phase differences between the cos $[\omega_H - \omega_L]$ term and the carrier terms are adjusted out. To complete the carrier extraction, the frequency $\omega_H - \omega_L$ is divided by an integer N and multiplied by the upper pilot tone. The difference frequency component is retained so that

$$\omega_c = \frac{(N-1)\omega_H + \omega_L}{N}. \tag{1.8.2}$$

For more details on this technique, the reader is referred to Lucky et al. (1968, pp. 184–186) and Holzman and Lawless (1970).

Today, modems rely heavily on suppressed carrier modulation methods such as AMVSB-SC, QAM, and PSK, which have no separate carrier component available. Of course, these modulation methods require a coherent reference for successful operation, and hence the carrier must be regenerated from the received data-carrying signal. There are a variety of techniques for accomplishing this, depending somewhat on the type of modulation being employed. Here we briefly discuss three of these techniques: the squaring loop, the Costas loop, and the Mth power method.

Consider a noise-free, generic AMDSB-SC signal given by

$$s(t) = m(t)\cos[\omega_c t + \phi], \tag{1.8.3}$$

from which we wish to extract a coherent reference signal. For the squaring loop shown in Fig. 1.14, $s(t)$ is passed through a square-law device that yields

$$s^2(t) = m^2(t)\left\{\frac{1}{2} + \frac{1}{2}\cos[2(\omega_c t + \phi)]\right\}. \tag{1.8.4}$$

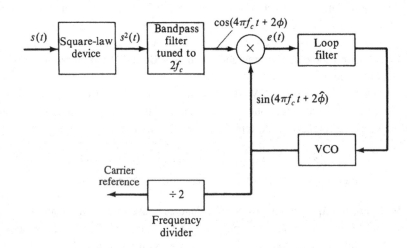

Fig. 1.14 Squaring loop for carrier recovery

   The purpose of the squaring operation is to reduce or eliminate the effects of the
modulating signal. To understand this, note that if $m(t) = \pm 1$, then $m^2(t) = 1$ for all
data sequences, while if $m(t)$ is multilevel, squaring eliminates phase changes and the
remaining amplitude variations can be removed by clipping. Only the double frequency
term is retained after bandpass filtering, and this component is fed to a tracking loop
(PLL) that locks on to the input frequency and phase. The VCO output is thus a sinusoid
at twice the frequency and phase of the received signal, but after frequency division by a
factor of 2, an accurate coherent reference is recovered.

   The Costas loop in Fig. 1.15 differs from the squaring loop in how it eliminates the
modulating signal effects and how it generates the input to the tracking loop. If the loop
is assumed to be locked on in frequency but with an inaccurate phase $\hat{\phi}$ as shown in
Fig. 1.15, then for $s(t)$ in Eq. (1.8.3),

$$y_1(t) = \frac{m(t)}{2} \cos(\phi - \hat{\phi}) + \frac{m(t)}{2} \cos\left(2\omega_c t + \phi + \hat{\phi}\right) \tag{1.8.5}$$

and

$$y_2(t) = \frac{m(t)}{2} \sin(\hat{\phi} - \phi) + \frac{m(t)}{2} \sin\left(2\omega_c t + \phi + \hat{\phi}\right). \tag{1.8.6}$$

Therefore, after filtering and multiplication, we have

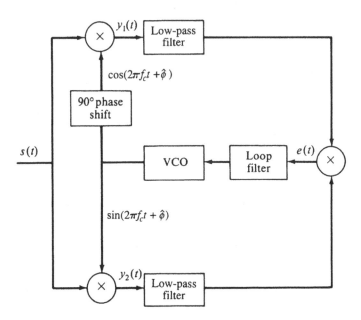

Fig. 1.15 Costas loop for carrier recovery

$$e(t) = \frac{m^2(t)}{8} \sin[2(\hat{\phi} - \phi)], \qquad (1.8.7)$$

which is then passed to the loop filter and VCO. Note that when $\hat{\phi} = \phi$, the output of the low-pass filter in the upper branch of Fig. 1.15 is the message or data sequence.

An M-phase (level) PSK waveform can be expressed as

$$s(t) = A_c \cos\left[\omega_c t + \frac{2\pi i}{M} + \phi\right], \qquad (1.8.8)$$

$i = 1, 2, \ldots, M$. We need to obtain from $s(t)$ a coherent reference signal of the form $\cos[\omega_c t + \phi]$. The Mth power method, which is similar to the squaring loop, is a technique for doing this. A block diagram of the Mth power method is shown in Fig. 1.16. If we pass $s(t)$ in Eq. (1.8.3) through the Mth power device in Fig. 1.16, we obtain for M even

$$s^M(t) = A_c^M \left\{ \frac{1}{2^M} \binom{M}{M/2} + \frac{1}{2^{M-1}} \left[\cos M\left(\omega_c t + \frac{2\pi i}{M} + \phi\right) \right.\right.$$
$$+ \binom{M}{1} \cos\left\{ (M-2)\left(\omega_c t + \frac{2\pi i}{M} + \phi\right) \right\}$$
$$\left.\left. + \cdots + \binom{M}{M/2 - 1} \cos 2\left(\omega_c t + \frac{2\pi i}{M} + \phi\right) \right] \right\}. \qquad (1.8.9)$$

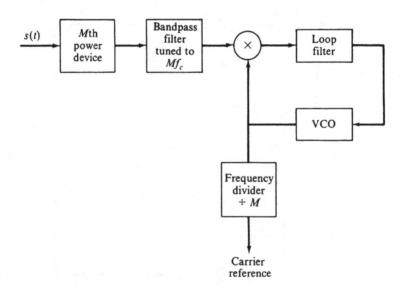

Fig. 1.16 Mth power method for carrier recovery

The bandpass filter tuned to $M\omega_c$ removes all terms except the one containing $\cos M(\omega_c t + 2\pi i/M + \phi) = \cos(M\omega_c t + M\phi)$. The frequency division by M gives the desired coherent reference signal. Additional details on carrier recovery are left to the literature (Franks 1980; Bhargava et al. 1981; Proakis 1989).

Symbol synchronization or timing extraction is usually accomplished by operations on the transmitted data sequence. One popular method relies on the threshold crossings of the equalized data sequence and on a highly stable crystal oscillator (Holzman and Lawless 1970). The crystal oscillator counts at a rate of about two orders of magnitude above the baud rate, and it effectively counts down from one sampling time to the next. The threshold crossings of the equalized baseband data sequence are used to center the sampling time on the open eye. This is accomplished by what is called an "early-late" decision in every pulse interval. If a transition occurs during the first half of the pulse (symbol) interval, it is desired to delay the sampling time so that it will be centered on the open eye. To accomplish this, we can require an extra "count" or we can effectively delete one of the previous counting pulses, so that this pulse must be counted again. The latter method is used in practice. If a threshold crossing occurs in the last half of the pulse interval, we wish to sample sooner. Hence we add a counting pulse, so that the countdown will be completed sooner. This technique continually centers the sampling instant on the open eye by adjusting the countdown chain in fixed increments equal to the reciprocal of the crystal oscillator counting rate. Accuracies of less than 1% of the baud rate are accomplished easily.

A very common method for symbol synchronization or clock recovery has much the same structure as the squaring loop for carrier recovery in Fig. 1.14. A block diagram of a general timing recovery loop is shown in Fig. 1.17. The demodulated symbol sequence is passed through a prefilter and then to a squarer or square-law device. As before, the squarer eliminates or reduces the effects of the particular data sequence. The squarer output is then sent to a BPF tuned to the symbol rate. The tracking loop extracts the

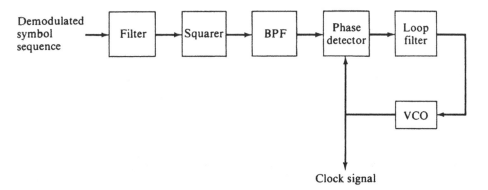

Fig. 1.17 General timing recovery method

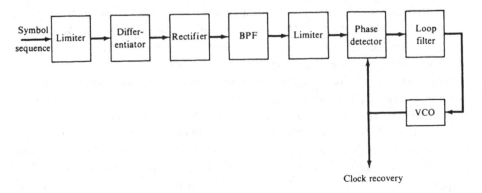

Fig. 1.18 Example clock recovery block diagram

clock signal from the BPF output. A common example of this approach is shown in block diagram form in Fig. 1.18. This latter clock recovery method is similar to that used in regenerative repeaters for pulse code modulation. Additional details are left to the references (Franks 1980; Bhargava et al. 1981; Proakis 1989).

1.9   Problems

1.1 Given a triangular time-domain pulse defined over $[-T/2, T/2]$ with maximum amplitude V, we know its Fourier transform from tables.
   (a) Sketch this Fourier transform pair.
   (b) Use the symmetry property of Fourier transforms to obtain a time-domain pulse with a triangular spectral content. Sketch this Fourier transform pair.
   (c) If we wish to transmit a sequence of pulses occurring every τ seconds using the time-domain pulse shape of part (b), what is the required bandwidth for zero intersymbol interference at sampling times? Compare your result to 100% roll-off raised cosine shaping.

1.2 Given the Fourier transform pair

$$f(t) = \begin{cases} V\cos\frac{\pi t}{T}, & |t| \le \frac{T}{2} \\ 0, & \text{otherwise,} \end{cases}$$

$$F(\omega) = \mathcal{F}\{f(t)\} = 2\pi V T \frac{\cos(\omega T/2)}{\pi^2 - \omega^2 T^2},$$

repeat Problem 1.1 for this function and its Fourier transform.

1.3 The input to a receiver is a sequence of positive and negative pulses represented by

$$r(t) = \sum_{n=-N}^{N} b_n p(t - n\tau),$$

where the b_ns are ± 1 and $p(t)$ is given by Eq. (1.2.3). To recover b_0, we wish to sample $y(t)$ at $t = 0$. Assume, instead, that due to timing jitter, we are in error by a small positive amount, say ε, so

$$r(\varepsilon) = \sum_{n=-N}^{N} b_n p(\varepsilon - n\tau).$$

Assume that $\varepsilon < \tau$ and show that as $N \to \infty$ there is a sequence for which $r(\varepsilon)$ diverges.

1.4 Verify that $100\,\alpha\%$ raised cosine pulse shaping satisfies Eq. (1.2.5). Use any of the three available expressions for $P(\omega)$: Eqs. (1.2.7), (1.2.9), or (1.2.10).

1.5 Verify that $100\,\alpha\%$ raised cosine pulse shaping satisfies Eq. (1.2.6). Use any of the three available expressions for $P(\omega)$: Eqs. (1.2.7), (1.2.9), or (1.2.10).

1.6 Rewrite each of the raised cosine spectral shaping expressions in Eqs. (1.2.7), (1.2.9), and (1.2.10) in terms of the transmitted data rate R, thereby obtaining explicit relationships between the data rate and required bandwidth.

1.7 Using a unit amplitude, 50% roll-off raised cosine pulse to indicate a 1 and no pulse to indicate a 0, sketch the pulse sequence corresponding to the binary message sequence 11001.

1.8 Starting with Eq. (1.2.10), demonstrate the equivalence of Eqs. (1.2.7), (1.2.9), and (1.2.10).

1.9 Given a fixed available bandwidth of 4000 Hz:
   (a) What is the maximum possible transmitted symbol rate using 100% roll-off raised cosine pulses?
   (b) What is the maximum possible transmitted symbol rate using 50% roll-off raised cosine pulses? Compare parts (a) and (b).

1.10 If we generate $100\alpha\%$ roll-off raised cosine pulses starting with rectangular pulses rather than impulses, to maintain the raised cosine shaping of the transmitted pulse, we must modify $P(\omega)$ to account for the rectangular pulse spectral content.
   (a) If unit-amplitude rectangular pulses of width Δ are to be used to generate $100\alpha\%$ roll-off raised cosine pulses, what should be the transfer function of the shaping filter?
   (b) It is desired to generate full cosine roll-off ($\alpha = 1$) raised cosine pulses from rectangular pulses one-tenth of the symbol interval in width. If the symbol rate is 2400 symbols/s, what is the required shaping filter frequency response?

1.11 A pulse-shaping function that satisfies Nyquist's third criterion is

$$P(\omega) = \begin{cases} \frac{\omega\tau/2}{\sin(\omega\tau/2)}, & 0 \le |\omega| \le \frac{\pi}{\tau} \\ 0, & |\omega| > \frac{\pi}{\tau}, \end{cases}$$

which has the impulse response

$$p(t) = \frac{1}{\pi} \int_0^{\pi/\tau} \frac{\omega\tau/2}{\sin(\omega\tau/2)} \cos \omega t \, d\omega.$$

Show that

$$\int_{(2n-1)\tau/2}^{(2n+1)\tau/2} p(t) dt = \begin{cases} 1, & n = 0 \\ 0, & n \ne 0, \end{cases}$$

which demonstrates that $p(t)$ has zero area for every symbol interval other than the one allocated to it, and further, that the response within this interval is directly proportional to the input impulse weight.

1.12 From Fig. 1.4, show that the duobinary impulse response can be written as

$$p(t) = \frac{\sin(\pi t/\tau)}{\pi t/\tau} + \frac{\sin[\pi(t-\tau)/\tau]}{[\pi(t-\tau)/\tau]}.$$

Sketch this $p(t)$, thereby verifying the impulse response waveform in Fig. 1.3.

1.13 For the $\{a_k\}$ sequence in Table 1.1, define a different $\{q_k\}$ sequence by the mapping

$$\begin{array}{cc} a_k & q_k \\ 0 \Rightarrow & -1 \\ 1 \Rightarrow & 1 \end{array}$$

Use Eq. (1.3.3) to represent this new $\{q_k\}$ sequence without precoding and develop a decoding rule. By assuming that s_3 is erroneously received, demonstrate that errors propagate in this scheme.

1.14 Sketch $s(t)$ for Table 1.1 and verify the values listed for the sequence $\{s_k\}$.

1.15 It is desired to transmit the binary sequence $\{a_k\}$ given by 01011010001 using duobinary signaling with precoding. Letting $b_0 = 1$, create an encoding table analogous to Table 1.1, listing the sequences $\{b_k\}$, $\{q_k\}$, and $\{s_k\}$. Show that an error in s_4 generates only a single decoding error.

1.16 For the input data sequence $\{a_k\}$ given by 01011010001, use class 4 partial response signaling with precoding and create an encoding table analogous to Table 1.2. Let $b_0 = 0$ and $b_{-1} = 1$. Demonstrate that if s_3 is received incorrectly, no error propagation occurs.

1.17 Using Eq. (1.3.9), sketch $s(t)$ corresponding to Table 1.2.

1.18 Use Fig. 1.4 and Eqs. (1.3.15)–(1.3.17) to verify Eqs. (1.3.1), (1.3.3), (1.3.4), and (1.3.6) for duobinary and Eqs. (1.3.7), (1.3.11), (1.3.13), and (1.3.14) for class 4 partial response.

1.19 Let $C_0 = -1$, $C_1 = 0$, $C_2 = 2$, $C_3 = 0$, and $C_4 = -1$ (all other C's zero) in Fig. 1.4, and show that

$$|P(\omega)| = 4\tau \sin^2 \omega\tau, \quad |\omega| \leq \frac{\pi}{\tau}$$

with $|P(\omega)| = 0$ otherwise. Sketch the corresponding impulse response of this shaping filter. This spectral shaping has been denoted class 5 partial response by Kretzmer (1966).

1.20 Double-sideband-suppressed carrier amplitude modulation (AMDSB-SC) is to be used to transmit data over a bandpass channel between 200 and 3200 Hz.

   (a) If the signaling rate (symbol rate) is 2000 symbols/s, what is the maximum raised cosine spectral shaping roll-off that can be used?

   (b) If 60% roll-off raised cosine pulses are used, what is the maximum possible signaling rate?

1.21 It is desired to transmit data at 9600 bits/s over a bandpass channel from 200 to 3200 Hz. If 50% roll-off raised cosine pulses are to be used, compare the symbol (baud) rate and the number of bits/symbol required by AMDSB-SC and AMSSB-SC modulation methods. Discuss the weaknesses of each system.

1.22 Only the range of frequencies from 500 to 3000 Hz are usable on a particular telephone channel. If 50% roll-off raised cosine pulse shaping is used and two levels per symbol are transmitted, calculate the maximum bit rate possible using:

   (a) AMVSB-SC with 1.2 times the baseband bandwidth.

   (b) Quadrature amplitude modulation.

1.23 Sketch the eye pattern for a data transmission system that sends eight levels in the amplitude range $[-V, V]$ using 100% roll-off raised cosine pulse shaping. Assume that the transmitted levels are $\pm V, \pm 5V/7, \pm 3V/7, +1V/7$, and sketch at least two pulse intervals and three sampling instants.

1.24 Sketch a three-level eye pattern for the pulses with quadratic delay distortion shown in Fig. 1.6. Use the pulses with the maximum distortion shown.

1.25 The input to a three-tap equalizer is $y_{-1} = 0.1$, $y_0 = 1.2$, $y_1 = -0.2$, all other $y_i = 0$. The desired equalizer output is $s_0 = 1$, all other $s_1 = 0$.
   (a) Choose the equalizer tap gains to produce $s_{-1} = 0$, $s_0 = 1$, $s_1 = 0$.
   (b) Choose the tap gains to minimize the MSE between the equalizer output and the desired values.
   (c) Find the output sequences for the equalizers in parts (a) and (b).
   (d) Calculate the MSE for parts (a) and (b). Compare the two values and discuss your results.

1.26 The input to a three-tap equalizer is $y_{-1} = 1/4$, $y_0 = 1$, $y_1 = 1/2$, all other $y_i = 0$. If the equalizer tap gains are $c_{-1} = -1/3$, $c_0 = 1$, and $c_1 = -2/3$, use polynomial multiplication to find the output sequence $\{s_k\}$ for all k.

1.27 The input to a three-tap equalizer is $y_{-1} = -\frac{1}{4}$, $y_0 = 1$, $y_1 = \frac{1}{4}$, y_0, for all other i. The desired output sequence is $s_0 = 1$, $s_i = 0$ for all other i. Find the equalizer tap gains to minimize the mean-squared error between the output and the desired values.

1.28 For the input sequence s_1 and initial shift register contents in Table 1.3, use the feedforward device as the scrambler and the feedback connection as the unscrambler.

1.29 In Table 1.4, change the third bit in the s_2' sequence and compute s_3. To illustrate the differences in error propagation, flip the third bit in the unscrambler input in Problem 1.28. Compare the results.

1.30 For the scrambler/unscrambler shown in the figure below, compute the scrambler output s_2 and the unscrambler output s_3 for the input sequence $s_1 = 011010011100$ and all zero initial shift register contents. Assume that there are no channel errors.

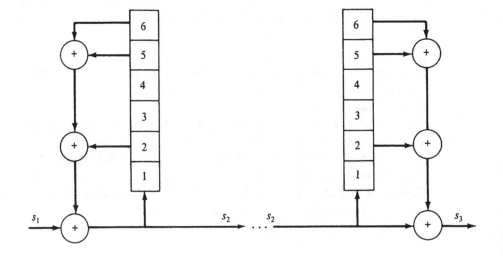

1.31 Consider a linear, feedback connection, four-stage shift register. Assume that the initial shift register contents are all zero and that the input sequence is a 1 at the first time instant followed by all zeros. Find the output sequence corresponding to each possible feedback connection. Specify the connection that yields a maximal-length sequence.

Noise in Digital Communications

2.1 Introduction

In this chapter we evaluate the performance of the data transmission and digital communication systems developed in Chap. 1 when the transmitted signals are disturbed by random noise. The transmitted waveforms in Chap. 1 can be quite different. The original digital message sequence can be transmitted using analog modulation methods, such as AMVSB-SC, FSK, PSK, and DPSK, or the transmitted signals can be discrete-amplitude "baseband" pulses. Despite these seemingly drastic physical differences, it turns out that we can express these transmitted signals in a common analytical framework, called signal space, which is developed in Sect. 2.2. Now, one of the reasons that the transmitted signals can be so different is that the channels over which the digital information is to be transmitted are different; that is, the channel can be a narrowband, bandpass channel, or a wideband baseband or low-pass channel.

Because these channels have such different physical characteristics, one feels intuitively that both the deterministic and random distortion present on these two channels would be quite dissimilar. This is indeed the case, but we are fortunate in that one channel model plays a dominant role in the design and analysis of a major portion of the existing telecommunications systems. This particular channel, the additive white Gaussian noise channel, is examined in Sect. 2.3, and there we find that not only does this model have wide applicability, but it is also tractable analytically—a fortuitous situation indeed.

Once we have the analytical framework of Sects. 2.2 and 2.3, we are able in Sect. 2.4 to derive optimum receiver structures for a wide variety of communication systems and to compute the error probability for these optimum systems and other suboptimum communication systems in Sect. 2.5. Probability of error (or bit error rate, as it is sometimes called) is the basis for our performance evaluations in this chapter, but we also know that

© The Author(s), under exclusive license to Springer Nature Switzerland AG 2023
J. D. Gibson, *Digital Communications*, Synthesis Lectures on Communications,
https://doi.org/10.1007/978-3-031-19588-4_2

there are other physical constraints on transmitted power, bandwidth, and perhaps, complexity, which must be satisfied. Hence, in Sect. 2.6, we present communication system
performance comparisons that illustrate a few of the many trade-offs involved in communication system design. Finally, in Sect. 2.7 we develop receiver structures that are
optimum in the presence of intersymbol interference and additive white Gaussian noise.

2.2 Signal Space and the Gram–Schmidt Procedure

The key to analyzing the performance of many communication systems is to realize that
the signals used in these communication systems can be expressed and visualized geometrically. In fact, many signals in today's communication systems are expressible in only
two dimensions! Since most undergraduate electrical engineers are quite comfortable with
geometric concepts in three dimensions or less and have numerous years of experience
with two-dimensional geometry, we are very attracted to a geometrical formulation of the
problem.

Let us begin the development by trying to define a two-dimensional representation of a
set of time-domain waveforms $\{s_1(t), s_2(t), \ldots, s_N(t)\}$ over the time interval $0 \le t \le T$.
We know that in standard two-dimensional spaces, we have two orthogonal vectors, a
concept of the length of a vector, and the idea of the distance between two vectors or two
points in the plane. Hence, if we can define similar concepts in the time domain, we can
then set up a one-to-one correspondence between the time domain and two-dimensional
space. We know that two (real) time functions $f(t)$ and $g(t)$ are said to be orthogonal over
the interval $[0, T]$ if

$$\int_0^T f(t)g(t)\, dt = 0 \qquad\qquad (2.2.1)$$

for $f(t) \ne g(t)$. Thus we already have the needed concept of orthogonality. We obtain
the concepts of the length of a signal and the distance between two signals by defining
the energy of a signal

$$\int_0^T f^2(t)\, dt = E_f \qquad\qquad (2.2.2)$$

as its length squared, and the energy of the difference between two signals given by

$$\int_0^T [f(t) - g(t)]^2 dt = d_{fg}^2 \qquad\qquad (2.2.3)$$

as the distance between the two signals squared. Thus the geometrical length corresponding to the waveform $f(t)$ is $\sqrt{E_f}$ from Eq. (2.2.2), and the geometrical distance between $f(t)$ and $g(t)$ is d_{fg} from Eq. (2.2.3). As a final step, we should define a pair of orthonormal functions to correspond to unit vectors in the two-dimensional space. To do this, we simply let

$$\varphi_1(t) = \frac{f(t)}{\sqrt{E_f}} \quad \text{and} \quad \varphi_2(t) = \frac{g(t)}{\sqrt{E_g}}, \tag{2.2.4}$$

so that

$$\int_0^T \varphi_i(t)\varphi_j(t)\, dt = \begin{cases} 0, \ i \neq j \\ 1, \ i = j. \end{cases} \tag{2.2.5}$$

Now, if we assume that the set of N signals $\{s_1(t), s_2(t), \ldots, s_N(t)\}$ can be expressed as a linear combination of the orthonormal waveforms $\varphi_1(t)$ and $\varphi_2(t)$, we can write

$$\begin{aligned} s_1(t) &= s_{11}\varphi_1(t) + s_{12}\varphi_2(t) \\ s_2(t) &= s_{21}\varphi_1(t) + s_{22}\varphi_2(t) \\ &\vdots \\ s_N(t) &= s_{N1}\varphi_1(t) + s_{N2}\varphi_2(t) \end{aligned} \tag{2.2.6}$$

Defining the column vectors

$$\varphi_1 = \begin{bmatrix} 1 \\ 0 \end{bmatrix} \quad \text{and} \quad \varphi_2 = \begin{bmatrix} 0 \\ 1 \end{bmatrix}, \tag{2.2.7}$$

we see that we can rewrite the set of equations in Eq. (2.2.6) as

$$\begin{aligned} \mathbf{s_1} &= s_{11}\varphi_1 + s_{12}\varphi_2 \\ \mathbf{s_2} &= s_{21}\varphi_1 + s_{22}\varphi_2 \\ &\vdots \\ \mathbf{s_N} &= s_{N1}\varphi_1 + s_{N2}\varphi_2 \end{aligned} \tag{2.2.8}$$

or as row vectors

$$\mathbf{s_1} = \begin{bmatrix} s_{11} & s_{12} \end{bmatrix}$$

$$\mathbf{s_2} = \begin{bmatrix} s_{21} & s_{22} \end{bmatrix} \tag{2.2.9}$$

$$\vdots$$

$$\mathbf{s_N} = \begin{bmatrix} s_{N1} & s_{N2} \end{bmatrix}.$$

In the set of equations denoted by Eq. (2.2.9), it is clear that the first component of the row vector is in the φ_1 direction and the second component is in the φ_2 direction, and we now have a correspondence between the physical, time-domain signal set $\{s_i(t), i = 1, 2, \ldots, N\}$ and points in a two-dimensional space. However, to complete the correspondence, we must find expressions in two-dimensional space analogous to Eqs. (2.2.2), (2.2.3), and (2.2.5). To do this, we consider the inner product of any two signals in the set given by

$$
\begin{aligned}
\int_0^T s_i(t)s_j(t)dt &= \int_0^T \sum_{n=1}^2 \sum_{m=1}^2 s_{in}s_{jm}\varphi_n(t)\varphi_m(t)dt \\
&= \sum_{n=1}^2 \sum_{m=1}^2 s_{in}s_{jm} \int_0^T \varphi_n(t)\varphi_m(t)dt \\
&= \sum_{n=1}^2 s_{in}s_{jn} = [s_{i1}s_{i2}]\begin{bmatrix} s_{j1} \\ s_{j2} \end{bmatrix} \\
&\triangleq \mathbf{s}_i \cdot \mathbf{s}_j,
\end{aligned}
\tag{2.2.10}
$$

which is a form of Parseval's relations. Equation (2.2.10) is a relationship between the inner product in the time domain and the inner product in our two-dimensional signal space. We consider

$$
\begin{aligned}
\int_0^T \varphi_i(t)\varphi_j(t)dt &= \varphi_i \cdot \varphi_j \\
&= \begin{cases} \begin{bmatrix} 1 & 0 \end{bmatrix}\begin{bmatrix} 0 \\ 1 \end{bmatrix} = 0, & i \neq j \\[2ex] \begin{bmatrix} 1 & 0 \end{bmatrix}\begin{bmatrix} 1 \\ 0 \end{bmatrix} = 1 \text{ or } \begin{bmatrix} 0 & 1 \end{bmatrix}\begin{bmatrix} 0 \\ 1 \end{bmatrix} = 1, & i = j, \end{cases}
\end{aligned}
\tag{2.2.11}
$$

which is the same as Eq. (2.2.5). An expression for the energy of any signal in the space is obtained by letting $i = j$ in Eq. (2.2.10), which yields

$$
\int_0^T s_i^2(t)dt = \sum_{n=1}^N s_{in}^2 = \mathbf{s}_i \cdot \mathbf{s}_i,
\tag{2.2.12}
$$

and distance between any two points (vectors) in the space can be obtained straightforwardly from Eq. (2.2.3) as

$$\int_0^T \left[s_i(t) - s_j(t)\right]^2 dt = \int_0^T \left[\sum_{n=1}^N (s_{in} - s_{jn})\varphi_n(t)\right]^2 dt$$

$$= \sum_{n=1}^N \sum_{m=1}^N (s_{in} - s_{jn})(s_{im} - s_{jm}) \int_0^T \varphi_n(t)\varphi_m(t)dt$$

$$= \sum_{n=1}^N (s_{in} - s_{jn})^2$$

$$\triangleq d_{ij}^2. \tag{2.2.13}$$

The correspondence between the original set of time-domain signals and our two-dimensional space, called a *signal space*, is now complete. For any $s_i(t)$, we obtain a vector (point) in the space from Eqs. (2.2.6)–(2.2.9) that has the length (distance from the origin) given by the square root of Eq. (2.2.12). Finally, the distance between any two vectors in the space is d_{ij}, which is available from Eq. (2.2.13).

Example 2.2.1 A very important example for actual applications is the set of multilevel quadrature amplitude-modulated signals. In this case, we have $\varphi_1(t) = \sqrt{2/T} \cos \omega_c t$ and $\varphi_2(t) = \sqrt{2/T} \sin \omega_c t$ as our orthonormal waveforms for $0 \le t \le T$. So that we can easily enumerate all the signals involved, let us consider initially $N = 4$ with the possible amplitudes of $\pm A_c$ in each direction. Therefore, we have four waveforms defined over $[0, T]$ as

$$s_1(t) = A_c\varphi_1(t) + A_c\varphi_2(t)$$

$$= A_c\sqrt{\frac{2}{T}} \cos \omega_c t + A_c\sqrt{\frac{2}{T}} \sin \omega_c t$$

$$s_2(t) = A_c\varphi_1(t) - A_c\varphi_2(t)$$

$$= A_c\sqrt{\frac{2}{T}} \cos \omega_c t - A_c\sqrt{\frac{2}{T}} \sin \omega_c t$$

$$s_3(t) = -A_c\varphi_1(t) + A_c\varphi_2(t)$$

$$= -A_c\sqrt{\frac{2}{T}} \cos \omega_c t + A_c\sqrt{\frac{2}{T}} \sin \omega_c t$$

$$s_4(t) = -A_c\varphi_1(t) - A_c\varphi_2(t)$$

$$= -A_c\sqrt{\frac{2}{T}} \cos \omega_c t - A_c\sqrt{\frac{2}{T}} \sin \omega_c t. \tag{2.2.14}$$

Each of these signals has equal energy given by

$$\int_0^T s_i^2(t)\, dt = 2A_c^2 \triangleq E_s. \tag{2.2.15}$$

We can represent this set of signals by vectors in two-dimensional signal space as

$$\mathbf{s}_1 = \begin{bmatrix} A_c & A_c \end{bmatrix}$$

$$\mathbf{s}_2 = \begin{bmatrix} A_c & -A_c \end{bmatrix}$$

$$\mathbf{s}_3 = \begin{bmatrix} -A_c & A_c \end{bmatrix}$$

$$\mathbf{s}_4 = \begin{bmatrix} -A_c & -A_c \end{bmatrix}, \tag{2.2.16}$$

where from Eq. (2.2.12),

$$\mathbf{s}_i \cdot \mathbf{s}_i = 2A_c^2. \tag{2.2.17}$$

A plot of the signal set in Eq. (2.2.16) is shown in Fig. 2.1 and is called a *signal space diagram*. Note that we must know the correspondence between the φ_i and the $\varphi_i(t)$ to go back to the physically meaningful time domain. Furthermore, the points in Fig. 2.1 can represent a wide variety of waveforms, depending on this correspondence.

The distance from the origin to each signal point is obvious as given by Eqs. (2.2.15) and (2.2.17). Furthermore, we can use Eq. (2.2.13) to calculate any d_{ij} of interest, say d_{23}, as

$$d_{23}^2 = \sum_{n=1}^2 (s_{2n} - s_{3n})^2 = (A_c - (-A_c))^2 + (-A_c - A_c)^2 = 8A_c^2,$$

so $d_{23} = 2\sqrt{2}A_c$ which is obviously true by inspection of Fig. 2.1.

Fig. 2.1 Signal space diagram for four-level QAM in Example 2.2.1 (see also Example 2.2.3)

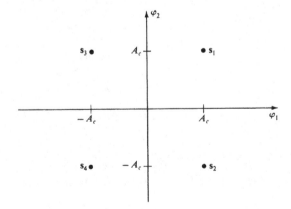

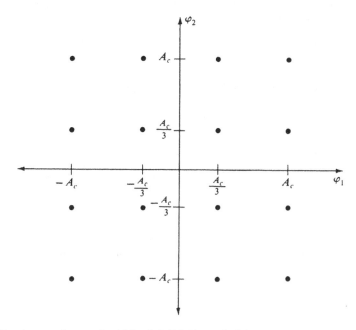

Fig. 2.2 Signal space diagram for 16-level QAM (Example 2.2.1)

If we retain the same orthonormal time functions $\varphi_1(t)$ and $\varphi_2(t)$ and now consider the case where we have four possible amplitudes in each direction, namely $\pm A_c/3, \pm A_c$, then we have $N = 16$, or what is usually called 16-level QAM. We leave it to the reader to write out the set of time-domain waveforms $\{s_i(t), i = 1, 2, \ldots, 16\}$ analogous to Eq. (2.2.16), but the signal space corresponding to 16-level QAM is shown in Fig. 2.2. Other details concerning the signal space representation of 16-level QAM are left to the problems.

Signal space diagrams are also often called *signal constellations*. To illustrate the generality of signal space diagrams further, we consider two additional examples.

Example 2.2.2 Phase shift keying (PSK) modulation has been, and continues to be, an important analog modulation technique for transmitting digital information. To begin, we consider binary PSK (BPSK), which has the two transmitted signals

$$s_1(t) = A_c\sqrt{\frac{2}{T}}\cos\omega_c t \qquad (2.2.18a)$$

and

$$s_2(t) = A_c\sqrt{\frac{2}{T}}\cos\left[\omega_c t + 180^\circ\right]$$

Fig. 2.3 Signal space diagram
for BPSK in Example 2.2.2
(see also Example 2.2.3)

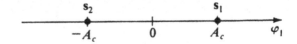

$$= -A_c\sqrt{\frac{2}{T}}\cos\omega_c t \qquad (2.2.18b)$$

for $0 \leq t \leq T$. If we let $\varphi(t) = \sqrt{2/T}\cos\omega_c t$, it is immediately evident that BPSK has a one-dimensional signal space representation as shown in Fig. 2.3.

If we now wish to consider four-phase PSK, we might use the four time-domain waveforms

$$s_i(t) = A_c\sqrt{\frac{2}{T}}\cos\left[\omega_c t + \frac{i\pi}{2}\right]$$

for $i = 1, 2, 3, 4$, and $0 \leq t \leq T$. Using a trigonometric identity, we obtain

$$s_i(t) = A_c\sqrt{\frac{2}{T}}\cos\frac{i\pi}{2}\cos\omega_c t - A_c\sqrt{\frac{2}{T}}\sin\frac{i\pi}{2}\sin\omega_c t. \qquad (2.2.19)$$

Letting $\varphi_1(t) = \sqrt{2/T}\cos\omega_c t$ and $\varphi_2(t) = \sqrt{2/T}\sin\omega_c t$, we can rewrite Eq. (2.2.19) as

$$s_i(t) = A_c\cos\frac{i\pi}{2}\varphi_1(t) - A_c\sin\frac{i\pi}{2}\varphi_2(t) \qquad (2.2.20)$$

for $i = 1, 2, 3, 4$. Comparing Eq. (2.2.20) for all i with Fig. 2.1, we see that except for a scaling by $\sqrt{2}$ and a renumbering of the signals, the signal space diagram for our four-phase PSK signals has the same form as Fig. 2.1, but it is rotated by $45°$. Depending on the application, this difference in rotation may or may not be of importance. This point is discussed in more detail in later sections of this chapter.

Eight-phase PSK is an important practical modulation method that can have the transmitted waveforms

$$s_i(t) = A_c\sqrt{\frac{2}{T}}\cos\left[\omega_c t - \frac{i\pi}{4}\right] \qquad (2.2.21)$$

for $i = 1, 2, ..., 8$, and $0 \leq t \leq T$. The signal space diagram for this signal set is shown in Fig. 2.4 and can be easily surmised by using a trig identity on Eq. (2.2.21).

Example 2.2.3 For this example we now consider the two orthonormal waveforms over $0 \leq t \leq T$:

$$\varphi_1(t) = \begin{cases} \sqrt{\frac{2}{T}}, & 0 \leq t \leq \frac{T}{2} \\ 0, & \frac{T}{2} < t \leq T \end{cases} \qquad (2.2.22)$$

Fig. 2.4 Signal space diagram
for eight-phase PSK in
Example 2.2.2

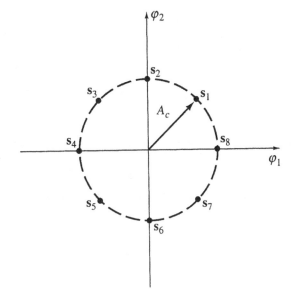

and

$$\varphi_2(t) = \begin{cases} 0, & 0 \leq t \leq \frac{T}{2} \\ \sqrt{\frac{2}{T}}, & \frac{T}{2} < t \leq T, \end{cases} \tag{2.2.23}$$

where both waveforms are zero outside $[0, T]$. Suppose now that we form the binary signal set (over $0 \leq t \leq T/2$)

$$s_1(t) = A_c \varphi_1(t) \tag{2.2.24a}$$

and

$$s_2(t) = -A_c \varphi_1(t). \tag{2.2.24b}$$

Then the signal space diagram for Eqs. (2.2.24a,b) is the same as in Fig. 2.3. Equations (2.2.24a,b) represent what are often called *binary antipodal signals*.

If we now let $N = 4$ with

$$\begin{aligned}
s_1(t) &= A_c \varphi_1(t) + A_c \varphi_2(t) \\
s_2(t) &= A_c \varphi_1(t) - A_c \varphi_2(t) \\
s_3(t) &= -A_c \varphi_1(t) + A_c \varphi_2(t) \\
s_4(t) &= -A_c \varphi_1(t) - A_c \varphi_2(t)
\end{aligned} \tag{2.2.25}$$

we discern immediately that the appropriate signal space diagram is identical to the one sketched in Fig. 2.1. If we compare the signal sets in Eqs. (2.2.14) and (2.2.25), this result is not surprising and again points out that a particular signal constellation can represent a variety of physical waveforms depending on the $\varphi_1(t)$ and $\varphi_2(t)$ functions. Can you specify a set of signals using Eqs. (2.2.22) and (2.2.23) that have the signal space diagram in Fig. 2.2? See Problem 2.6.

We have considered only two-dimensional signal spaces thus far (because of their physical importance and because of our geometric intuition); however, it is interesting to inquire as to the generality of the approach. That is, if we are presented with a set of N waveforms $\{s_i(t), i = 1, 2, \ldots, N\}$ defined over $-\infty < t < \infty$ with $\int_{-\infty}^{\infty} s_i^2(t)\,dt < \infty$, can we express each of these signals as a weighted linear combination of orthonormal waveforms

$$s_i(t) = \sum_{j=1}^{M} s_{ij}\varphi_j(t), \tag{2.2.26}$$

where $M \leq N$ and the set $\{\varphi_j(t), j = 1, 2, \ldots, M\}$ satisfies Eq. (2.2.5)? To accomplish an expansion of the form in Eq. (2.2.26), we must first be able to determine a suitable set of orthonormal waveforms. There is a well-defined technique for doing this called the Gram–Schmidt orthogonalization procedure.

Gram–Schmidt Orthogonalization

Here we define a straightforward procedure for generating $M \leq N$ orthonormal functions given any set of N finite-energy waveforms defined over $-\infty < t < \infty$. It is presented in a step-by-step fashion.

Step 1. For $s_1(t) \not\equiv 0$, find $E_1 = \int_{-\infty}^{\infty} s_1^2(t)\,dt$ and set

$$\varphi_1(t) = \frac{1}{\sqrt{E_1}} s_1(t). \tag{2.2.27}$$

Note that in the form of Eq. (2.2.26),

$$s_1(t) = s_{11}\varphi_1(t) = \sqrt{E_1}\varphi_1(t) \tag{2.2.28}$$

and further that we can find s_{11} from $s_1(t)$ as

$$s_{11} = \int_{-\infty}^{\infty} s_1(t)\varphi_1(t)\,dt. \tag{2.2.29}$$

Step 2. Form

$$f_2(t) = s_2(t) - s_{21}\varphi_1(t), \qquad (2.2.30)$$

where

$$s_{21} = \int\limits_{-\infty}^{\infty} s_2(t)\varphi_1(t)\, dt, \qquad (2.2.31)$$

so that $f_2(t)$ is orthogonal to $\varphi_1(t)$, that is,

$$\int\limits_{-\infty}^{\infty} f_2(t)\varphi_1(t)\, dt = \int\limits_{-\infty}^{\infty} [s_2(t) - s_{21}\varphi_1(t)]\varphi_1(t)\, dt = 0. \qquad (2.2.32)$$

Find $\int_{-\infty}^{\infty} f_2^2(t)\, dt$ and define

$$\varphi_2(t) = \frac{f_2(t)}{\sqrt{\int_{-\infty}^{\infty} f_2^2(t)\, dt}}. \qquad (2.2.33)$$

Note that

$$s_{22} = \int\limits_{-\infty}^{\infty} f_2(t)\varphi_2(t)\, dt = \sqrt{\int_{-\infty}^{\infty} f_2^2(t)\, dt}, \qquad (2.2.34)$$

so that

$$s_2(t) = s_{21}\varphi_1(t) + s_{22}\varphi_2(t). \qquad (2.2.35)$$

Clearly, $\varphi_2(t)$ is orthogonal to $\varphi_1(t)$ from Eqs. (2.2.32) and (2.2.33).

$\vdots$

Step k: The general step in the procedure can be specified by assuming that we have
generated $m - 1$ orthonormal waveforms $\{\varphi_j(t),\ j = 1, 2, \ldots m - 1\}$ based on k
$-\ 1$ of the original signals $\{s_i(t),\ i = 1, 2, \ldots, k - 1\}$. We will have $m - 1 \le k$
$-\ 1$, with strict inequality if at one or more of the preceding steps the $f_i(t) = 0$.
This occurrence implies that the original signal set is not linearly independent.
Continuing the process, we form

$$f_k(t) = s_k(t) - \sum_{j=1}^{m-1} s_{kj}\varphi_j(t), \qquad (2.2.36)$$

where the coefficients are

$$s_{kj} = \int_{-\infty}^{\infty} s_k(t) \varphi_j(t) \, dt. \tag{2.2.37}$$

If $f_k(t) \neq 0$, find $\int_{-\infty}^{\infty} f_k^2(t) \, dt$ and define

$$\varphi_m(t) = \frac{f_k(t)}{\sqrt{\int_{-\infty}^{\infty} f_k^2(t) dt}}. \tag{2.2.38}$$

Clearly, $f_k(t)$ is orthogonal to each member of the set, $\{\varphi_j(t), j = 1, 2, \ldots, m - 1\}$, so the set of $\varphi_j(t)$s, including $\varphi_m(t)$, are orthonormal. We have $s_{kj}, j = 1, 2, \ldots, m - 1$, so to write an expression for $s_k(t)$ of the form of Eq. (2.2.26), we need s_{km}. However,

$$s_{km} = \int_{-\infty}^{\infty} \varphi_m(t) f_k(t) \, dt = \sqrt{\int_{-\infty}^{\infty} f_k^2(t) \, dt} \tag{2.2.39}$$

and the expansion is completed.

The process is repeated until all N original signals have been used. The final result is an expansion as in Eq. (2.2.26).

Example 2.2.4 To illustrate the Gram–Schmidt procedure, we begin with the four signals $\{s_i(t), i = 1, 2, 3, 4\}$ shown in Fig. 2.5. We follow the steps in the procedure as outlined previously.

Step 1. Calculate

$$E_1 = \int_{0}^{T/3} (1)^2 \, dt = \frac{T}{3},$$

so

$$\varphi_1(t) = \sqrt{\frac{3}{T}} s_1(t) \tag{2.2.40}$$

and

$$s_1(t) = s_{11}\varphi_1(t) = \sqrt{\frac{T}{3}} \varphi_1(t). \tag{2.2.41}$$

Step 2. Find

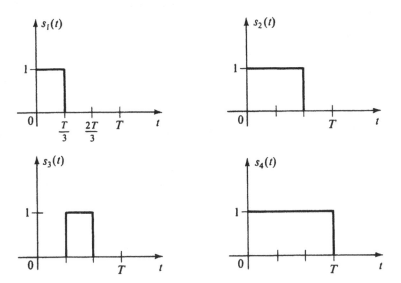

Fig. 2.5 Signal set for Example 2.2.4

$$s_{21} = \int_{-\infty}^{\infty} s_2(t)\varphi_1(t)\, dt = \int_0^{T/3} \sqrt{\frac{3}{T}}\, dt = \sqrt{\frac{T}{3}}, \qquad (2.2.42)$$

then form

$$f_2(t) = s_2(t) - s_{21}\varphi_1(t)$$
$$= s_2(t) - s_1(t). \qquad (2.2.43)$$

Now

$$\int_{-\infty}^{\infty} f_2^2(t)\, dt = \int_{T/3}^{2T/3} (1)^2\, dt = \frac{T}{3}, \qquad (2.2.44)$$

so we define

$$\varphi_2(t) = \sqrt{\frac{3}{T}}[s_2(t) - s_1(t)] \qquad (2.2.45)$$

and therefore

$$s_2(t) = s_{21}\varphi_1(t) + s_{22}\varphi_2(t)$$
$$= \sqrt{\frac{T}{3}}\varphi_1(t) + \sqrt{\frac{T}{3}}\varphi_2(t), \qquad (2.2.46)$$

since s_{22} is given by Eq. (2.2.34).

Step 3. Calculate

$$s_{31} = \int_{-\infty}^{\infty} s_3(t)\varphi_1(t)\, dt = 0, \qquad (2.2.47)$$

since $s_3(t)$ and $\varphi_1(t)$ do not overlap, and

$$s_{32} = \int_{-\infty}^{\infty} s_3(t)\varphi_2(t)\, dt = \int_{T/3}^{2T/3} \sqrt{\frac{3}{T}}\, dt = \sqrt{\frac{T}{3}}, \qquad (2.2.48)$$

and form

$$f_3(t) = s_3(t) - \sum_{j=1}^{2} s_{3j}\varphi_j(t)$$

$$= s_3(t) - \sqrt{\frac{T}{3}}\varphi_2(t)$$

$$= s_3(t) - \sqrt{\frac{T}{3}} \cdot \sqrt{\frac{3}{T}}[s_2(t) - s_1(t)]. \qquad (2.2.49)$$

But from Fig. 2.5, we observe that $s_2(t) - s_1(t) = s_3(t)$, so

$$f_3(t) = 0 \qquad (2.2.50)$$

and

$$s_3(t) = \sqrt{\frac{T}{3}}\varphi_2(t). \qquad (2.2.51)$$

From Eq. (2.2.50) we conclude that $s_3(t)$ is a weighted linear combination of $s_1(t)$ and $s_2(t)$, and therefore we proceed to the next step.

Step 4. Find

$$s_{41} = \int_{-\infty}^{\infty} s_4(t)\varphi_1(t)\, dt = \int_{0}^{T/3} \sqrt{\frac{3}{T}}\, dt = \sqrt{\frac{T}{3}} \qquad (2.2.52)$$

$$s_{42} = \int_{-\infty}^{\infty} s_4(t)\varphi_2(t)\, dt = \int_{T/3}^{2T/3} \sqrt{\frac{3}{T}}\, dt = \sqrt{\frac{T}{3}} \qquad (2.2.53)$$

and form

$$f_4(t) = s_4(t) - \sum_{j=1}^{2} s_{4j}\varphi_j(t)$$

$$= s_4(t) - s_1(t) - [s_2(t) - s_1(t)]$$

$$= \begin{cases} 1, & \text{for } \frac{2T}{3} \le t \le T \\ 0, & \text{otherwise.} \end{cases} \tag{2.2.54}$$

So

$$\int_{-\infty}^{\infty} f_4^2(t)\, dt = \frac{T}{3} \tag{2.2.55}$$

and

$$\varphi_3(t) = \sqrt{\frac{3}{T}} f_4(t). \tag{2.2.56}$$

Finally,

$$s_{43} = \sqrt{\frac{T}{3}} \tag{2.2.57}$$

and thus the desired orthonormal expansion is

$$s_4(t) = \sqrt{\frac{T}{3}}\, \varphi_1(t) + \sqrt{\frac{T}{3}}\, \varphi_2(t) + \sqrt{\frac{T}{3}}\, \varphi_3(t). \tag{2.2.58}$$

The orthonormal functions $\varphi_1(t)$, $\varphi_2(t)$, and $\varphi_3(t)$ are sketched in Fig. 2.6. Note that for this example $M = 3 < N = 4$.

The signal vector expressions for the $s_i(t)$ are

$$s_1 = \left[\sqrt{\frac{T}{3}}\ 0\ 0 \right]$$

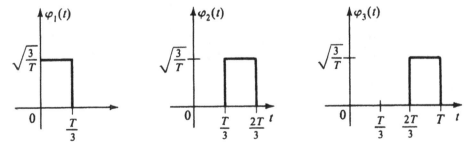

Fig. 2.6 Orthonormal functions for the signal set of Example 2.2.4

$$s_2 = \left[\sqrt{\tfrac{T}{3}} \ \sqrt{\tfrac{T}{3}} \ 0 \right]$$

$$s_3 = \left[0 \ \sqrt{\tfrac{T}{3}} \ 0 \right]$$

$$s_4 = \left[\sqrt{\tfrac{T}{3}} \ \sqrt{\tfrac{T}{3}} \ \sqrt{\tfrac{T}{3}} \right]. \tag{2.2.59}$$

Based on these vectors, we can compute intersignal distances, energies, and inner products in the three-dimensional signal space as needed.

There are several geometrically simple and practically important M-dimensional signal sets, where $M > 2$. We do not pursue these signal sets here; however, several problems are dedicated to the most common ones.

2.3   The Additive White Gaussian Noise Channel

We have discussed in Chap. 1 the effects of deterministic distortion and how we can compensate for this type of distortion by using equalization. We also noted in Chap. 1 that various kinds of random disturbances are present in communication systems, such as impulsive noise, abrupt phase changes (hits), and additive noise, just to name a few. One type of distortion evident in most communication systems can be accurately modeled as additive white Gaussian noise (AWGN). For some communications channels, such as the deep-space channel, AWGN is the principal random disturbance present, while for other channels, other random impairments (impulsive noise, for example) are a greater cause of error events. It is common to design systems such that they perform well in the presence of AWGN. There are several reasons usually given for this approach. First, disturbances such as impulsive noise have proven difficult to model. Second, when impulsive noise occurs, it is often of such a large amplitude that a system designed to prevent errors due to this phenomenon would be relatively inefficient when the impulsive noise is not present. Third, for high-speed data transmission, there are so many possible transmitted signals that AWGN does become a significant source of errors. Fourth, practical experience accumulated over many years indicates that communication systems designed to protect against AWGN also perform well in the presence of other disturbances. Fifth, the AWGN channel model is analytically tractable; that is, we can handle it mathematically. For all of the reasons just given, we spend the remainder of this chapter discussing the AWGN channel model and designing communication systems that perform well over such channels.

Given a transmitted time-domain waveform $s_i(t)$, we assume that this signal is contaminated by an additive disturbance $n(t)$, so that the waveform at the receiver input is

$$r(t) = s_i(t) + n(t), \tag{2.3.1}$$

where $n(t)$ is a zero-mean, white Gaussian noise process with power spectral density

$$S_n(\omega) = 2\pi \left(\frac{N_0}{2} \right) \quad \text{W/rad/s}, \quad -\infty < \omega < \infty. \tag{2.3.2}$$

Of course, we can represent the transmitted waveforms $s(t)$ as some vector in signal space, as discussed in Sect. 2.2, and we wonder if it is possible to represent $n(t)$, and hence $r(t)$, in the same signal space. Although we do not prove it here, such a representation is in fact possible, and we proceed directly to the vector space formulation of Eq. (2.3.1). If the transmitted waveform $s_i(t)$ can be expressed as

$$s_i(t) = \sum_{j=1}^{M} s_{ij} \varphi_j(t), \tag{2.3.3}$$

where

$$s_{ij} = \int_{-\infty}^{\infty} s_i(t)\varphi_j(t) \, dt, \tag{2.3.4}$$

then we have the vector representation

$$\mathbf{s}_i = \begin{bmatrix} s_{i1} & s_{i2} & \cdots & s_{iM} \end{bmatrix}. \tag{2.3.5}$$

The vector representation for the noise follows similarly as

$$\mathbf{n} = \begin{bmatrix} n_1 & n_2 & \cdots & n_M \end{bmatrix}, \tag{2.3.6}$$

where

$$n_j = \int_{-\infty}^{\infty} n(t)\varphi_j(t) \, dt, \tag{2.3.7}$$

so

$$\mathbf{r} = \begin{bmatrix} r_1 & r_2 & \cdots & r_M \end{bmatrix} = \mathbf{s}_i + \mathbf{n} \tag{2.3.8}$$

with

$$r_j = \int_{-\infty}^{\infty} r(t)\varphi_j(t) \, dt. \tag{2.3.9}$$

Based on the given conditions on $n(t)$, it can be demonstrated that the components of $\mathbf{n}$, namely the $\{n_j, j = 1, 2, \ldots, M\}$, are independent, identically distributed Gaussian

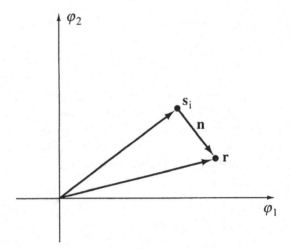

Fig. 2.7 Signal space AWGN channel model

random variables with zero mean and variance $\mathcal{N}_0/2$. The pdf of the random vector $\mathbf{n}$ is thus given by

$$f_{\mathbf{n}}(\mathbf{n}) = \frac{1}{[\pi \mathcal{N}_0]^{M/2}} e^{-\sum_{j=1}^{M} n_j^2/\mathcal{N}_0}. \tag{2.3.10}$$

A fact that will be important later is that the pdf in Eq. (2.3.10) is spherically symmetric, which means that the pdf depends only on the magnitude of the noise vector but not its direction.

Equation (2.3.8) thus expresses our signal space model for the received signal, which can be illustrated for two dimensions as shown in Fig. 2.7. Therefore, the additive noise vector causes the transmitted signal vector $\mathbf{s}_i$ to be observed at the receiver as another point in signal space, $\mathbf{r}$. The receiver must then make its decision as to which of the N signal vectors $\{\mathbf{s}_j, j = 1, 2, \ldots, N\}$ were transmitted based on the received vector $\mathbf{r}$ and knowledge of the noise vector pdf in Eq. (2.3.10).

2.4 Optimum Receivers

Using the signal space representations established in Sects. 2.2 and 2.3, we show in this section how to design receivers that are optimum in the sense that they minimize the probability of error, denoted $P[\mathcal{E}]$, for the given transmitted signal set and noise vector pdf. We find that these receivers have an elegant and intuitively pleasing interpretation in signal space. Before proceeding, however, let us consider a block diagram of the overall communication system of interest to us here.

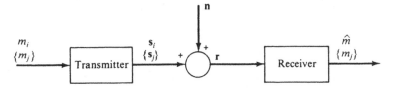

Fig. 2.8 Block diagram of the communication system

A block diagram of the communication system in terms of our vector space formulation is shown in Fig. 2.8. The input to the transmitter m_i is one of a set of N messages $\{m_j, j = 1, 2, \ldots, N\}$. The transmitter assigns a signal vector s_i (waveform) to the message to be sent from the set of possible transmitted vectors $\{s_j, j = 1, 2, \ldots, N\}$. The transmitted vector is then disturbed by the additive noise vector $\mathbf{n}$ so that the received vector is $\mathbf{r} = s_i + \mathbf{n}$. The receiver processes $\mathbf{r}$ to obtain the best possible estimate, $\hat{m}$, of which of the finite number of messages was being communicated. We now turn our attention to the design of the receiver.

For generality, we assume that we know the a priori probabilities of the individual messages, which we denote by $P[m_j]$. We desire to find a receiver that minimizes $P[\mathcal{E}]$, or equivalently, maximizes the probability of a correct decision, $P[\mathcal{C}]$. If we let $f_\mathbf{r}(\mathbf{r})$ denote the pdf of the received vector, the probability of a correct decision can be written as

$$P[\mathcal{C}] = \int_{-\infty}^{\infty} P[\mathcal{C}|\mathbf{r} = \alpha] f_\mathbf{r}(\alpha) d\alpha. \qquad (2.4.1)$$

Since $f_\mathbf{r}(\alpha) \geq 0$ and does not depend on the receiver, we see that we maximize $P[\mathcal{C}]$ if we maximize $P[\mathcal{C}|\mathbf{r} = \alpha]$. Now, if s_k is transmitted, then when the receiver chooses $\hat{m} = m_k$,

$$P[\mathcal{C}|\mathbf{r} = \alpha] = P[m_k|\mathbf{r} = \alpha]. \qquad (2.4.2)$$

Therefore, the receiver that maximizes $P[\mathcal{C}]$ sets $\hat{m}$ equal to that m_i which has the maximum a posteriori probability of occurrence $P[m_i|\mathbf{r} = \alpha]$.

Using Bayes' rule, we can expand the a posteriori message probabilities as

$$P[m_i|\mathbf{r} = \alpha] = \frac{f_{\mathbf{r}|m_i}(\alpha|m_i)P[m_i]}{f_\mathbf{r}(\alpha)}. \qquad (2.4.3)$$

With a little thought, we note that we need not retain the denominator on the right side of Eq. (2.4.3), since it is independent of the index i. Therefore, the optimum receiver sets $\hat{m} = m_k$ if (and only if) for all $i \neq k$,

$$f_{\mathbf{r}|m_k}(\alpha|m_k)P[m_k] > f_{\mathbf{r}|m_i}(\alpha|m_i)P[m_i]. \qquad (2.4.4)$$

Since conditioning on m_i is equivalent to conditioning on $\mathbf{s}_i$, the optimum receiver sets $\hat{m} = m_k$ if and only if for all $i \neq k$,

$$f_{\mathbf{r}|\mathbf{s}_k}(\alpha|\mathbf{s}_k)P[m_k] > f_{\mathbf{r}|\mathbf{s}_i}(\alpha|\mathbf{s}_i)P[m_i]. \qquad (2.4.5)$$

We can make this result much more explicit by using Eq. (2.3.10) to find the conditional pdfs $f_{\mathbf{r}|\mathbf{s}_i}(\alpha|\mathbf{s}_i)$. We know that $\mathbf{r} = \alpha = \mathbf{s}_i + \mathbf{n}$, so

$$f_{\mathbf{r}|\mathbf{s}_i}(\alpha|\mathbf{s}_i) = f_{\mathbf{n}}(\alpha - \mathbf{s}_i|\mathbf{s}_i). \qquad (2.4.6)$$

Assuming that the transmitted signal vector $\mathbf{s}_i$ and the noise are statistically independent, which is a physically reasonable assumption to make, we find that $f_{\mathbf{n}}(\alpha - \mathbf{s}_i|\mathbf{s}_i) = f_{\mathbf{n}}(\alpha - \mathbf{s}_i)$, so

$$f_{\mathbf{r}|\mathbf{s}_i}(\alpha|\mathbf{s}_i) = f_{\mathbf{n}}(\alpha - \mathbf{s}_i). \qquad (2.4.7)$$

Upon invoking Eq. (2.3.10) and noting the result in Eq. (2.4.7), the optimum receiver finds that i which maximizes the statistic

$$
\begin{aligned}
f_{\mathbf{r}|\mathbf{s}_i}(\alpha|\mathbf{s}_i)P[m_i] &= f_{\mathbf{n}}(\alpha - \mathbf{s}_i)P[m_i] \\
&= \frac{1}{[\pi\mathcal{N}_0]^{M/2}}e^{-\Sigma_{j=1}^{M}(\alpha_j - s_{ij})^2/\mathcal{N}_0}P[m_i]. \qquad (2.4.8)
\end{aligned}
$$

To simplify this, we drop the factor $[\pi\mathcal{N}_0]^{M/2}$, since it is independent of i, and take the natural logarithm of the remaining terms to obtain

$$-\sum_{j=1}^{M}\frac{(\alpha_j - s_{ij})^2}{\mathcal{N}_0} + \ln P[m_i]. \qquad (2.4.9)$$

Maximizing Eq. (2.4.9) is the same as minimizing the negative of the quantity, hence the optimum receiver sets $\hat{m} = m_k$ whenever

$$\sum_{j=1}^{M}(\alpha_j - s_{ij})^2 - \mathcal{N}_0 \ln P[m_i] \qquad (2.4.10)$$

is a minimum for $i = k$. We can also rewrite this result in the form of Eq. (2.4.5); that is, the optimum receiver sets $\hat{m} = m_k$ if and only if for all $i \neq k$,

$$\sum_{j=1}^{M}(\alpha_j - s_{kj})^2 - \mathcal{N}_0 \ln P[m_k] < \sum_{j=1}^{M}(\alpha_j - s_{ij})^2 - \mathcal{N}_0 \ln P[m_i]. \qquad (2.4.11)$$

Note that if the messages are all equally likely, then $P[m_i] = P[m_k]$ for all i, k, so Eq. (2.4.11) becomes

$$\sum_{j=1}^{M} (\alpha_j - s_{kj})^2 < \sum_{j=1}^{M} (\alpha_j - s_{ij})^2. \qquad (2.4.12)$$

Therefore, for equal a priori message probabilities, the receiver chooses $\hat{m} = m_k$ if and only if the received vector $\mathbf{r} = \alpha$ is closer to $\mathbf{s}_k$, in terms of Euclidean distance, than to any other $\mathbf{s}_i$, $i \neq k$.

For two dimensions, then, the optimum receiver is simply a partitioning of the two-dimensional signal space into regions, the points of which are closest to a given transmitted signal vector $\mathbf{s}_i$. These regions are called *decision regions*. We illustrate this general result by several examples.

Example 2.4.1 As our first illustration, we consider the transmission of four equally likely messages over an AWGN channel with power spectral density $\mathcal{N}_0/2$ W/Hz using QAM in Example 2.2.1. From Eq. (2.4.12), we see that the optimum decision regions are just the four quadrants; that is, the optimum receiver sets

$$\hat{m} = m_1 \text{ if } \mathbf{r} = \alpha \text{ is in the first quadrant,}$$

$$\hat{m} = m_3 \text{ if } \mathbf{r} = \alpha \text{ is in the second quadrant,}$$

$$\hat{m} = m_4 \text{ if } \mathbf{r} = \alpha \text{ is in the third quadrant,}$$

$$\hat{m} = m_2 \text{ if } \mathbf{r} = \alpha \text{ is in the fourth quadrant.}$$

The boundaries of the decision regions are the φ_1 and φ_2 axes.

As a second example, consider using the QAM signal set in Fig. 2.2 to send 16 equally likely messages over an AWGN channel. The resulting decision region boundaries are shown as dashed lines in Fig. 2.9 and include the φ_1 and φ_2 axes. Thus, if $\alpha = [\alpha_1 \ \alpha_2]$ is such that $\alpha_1 > 2A_c/3$ and $\alpha_2 > 2A_c/3$, the optimum receiver sets $\hat{m} = m_1$, since α is closest to $\mathbf{s}_1$ in the signal space. Also, if $0 < \alpha_1 < 2A_c/3$ and $0 < \alpha_2 < 2A_c/3$, then the optimum receiver sets $\hat{m} = m_6$. The optimum receiver thus has a simple geometric interpretation for these QAM signal sets.

Example 2.4.2 As a second example, we consider using PSK modulation to represent equally likely messages over an AWGN channel. For the binary case, we have the signals in Eqs. (2.2.18a) and (2.2.18b) with the signal space diagram in Fig. 2.3. There are only two decision regions for this case, and the decision boundary is the dashed line through 0 shown in Fig. 2.10. Thus, if $r = \alpha > 0$, $\hat{m} = m_1$ and if $r = \alpha < 0$, $\hat{m} = m_2$.

Similar to the other examples, we can sketch the optimum decision boundaries for the eight-phase PSK signal set in Fig. 2.4 as shown in Fig. 2.11. Of course, we are again assuming

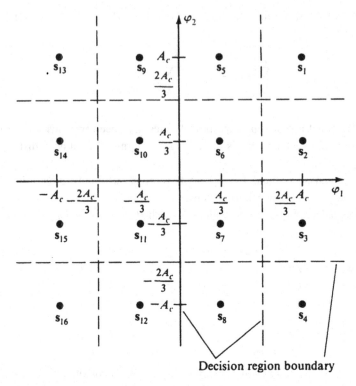

Fig. 2.9 Optimum decision regions for equally likely 16-level QAM (Example 2.4.1)

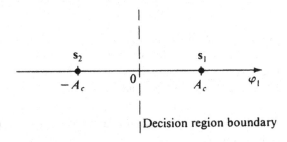

Fig. 2.10 Optimum decision
regions for BPSK (Example
2.4.2)

equally likely messages and an AWGN channel. Note that the decision regions in Fig. 2.11
are not as trivially written as inequalities on the components of α as in the QAM case.

Before concluding this example, let us return to the BPSK case and assume that $P[m_1] >
P[m_2]$. From Eq. (2.4.11), we see that the optimum receiver selects $\hat{m} = m_1$ if

$$\sum_{j=1}^{2} (\alpha_j - s_{1j})^2 - \mathcal{N}_0 \ln P[m_1] < \sum_{j=1}^{2} (\alpha_j - s_{2j})^2 - \mathcal{N}_0 \ln P[m_2] \qquad (2.4.13)$$

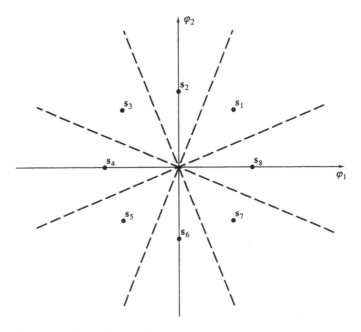

Fig. 2.11 Optimum decision regions for eight-phase PSK (Example 2.4.2)

and $\hat{m} = m_2$ if the inequality is reversed. We can write this optimum decision rule in a compact form as

$$\sum_{j=1}^{2} (\alpha_j - s_{1j})^2 - \mathcal{N}_0 \ln P[m_1] \underset{m_2}{\overset{m_1}{\lessgtr}} \sum_{j=1}^{2} (\alpha_j - s_{2j})^2 - \mathcal{N}_0 \ln P[m_2], \qquad (2.4.14)$$

where the labels on the inequality signs indicate the decision if the corresponding inequality is satisfied. Collecting terms involving a priori probabilities on the left side of Eq. (2.4.14) yields

$$\sum_{j=1}^{2} (\alpha_j - s_{1j})^2 - \mathcal{N}_0 \ln \frac{P[m_1]}{P[m_2]} \underset{m_2}{\overset{m_1}{\lessgtr}} \sum_{j=1}^{2} (\alpha_j - s_{2j})^2. \qquad (2.4.15)$$

Now, since $P[m_1] > P[m_2]$ and $\mathcal{N}_0 > 0$, the term $-\mathcal{N}_0 \ln(P[m_1]/P[m_2])$ is negative, so the boundary between the two decision regions is shifted away from s_1 nearer to s_2. This shifted boundary is illustrated in Fig. 2.12. Note that this result is intuitive, since m_1 is more likely to occur than m_2, and hence should be associated with an increased portion of signal space to minimize $P[\mathcal{E}]$.

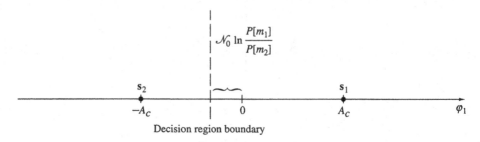

Fig. 2.12 Decision regions for BPSK with $P[m_1] > P[m_2]$ (Example 2.4.2)

Thus far we have only specified our receivers in terms of operations in signal space. We now present receiver implementations that perform the signal space operations. Note from Eq. (2.4.10) that the optimum receiver first calculates the $r_j = \alpha_j$ components according to Eq. (2.3.9) and then computes the quantity in Eq. (2.4.10) for all i. If we expand the first term in Eq. (2.4.10) (letting $r_j = \alpha_j$),

$$\sum_{j=1}^{M} (r_j - s_{ij})^2 = \sum_{j=1}^{M} r_j^2 - 2\sum_{j=1}^{M} r_j s_{ij} + \sum_{j=1}^{M} s_{ij}^2. \tag{2.4.16}$$

We see that the term $\sum r_j^2$ is independent of i, so that we can rewrite the optimum decision rule as choosing $\hat{m} = m_k$ to yield a minimum of

$$-2\sum_{j=1}^{M} r_j s_{ij} + \sum_{j=1}^{M} s_{ij}^2 - \mathcal{N}_0 \ln P[m_i] \tag{2.4.17}$$

for $i = k$. Equivalently, we can state the optimum decision rule as setting $\hat{m} = m_k$ if and only if the expression

$$\sum_{j=1}^{M} r_j s_{ij} + \frac{1}{2}\left[\mathcal{N}_0 \ln P(m_i) - \sum_{j=1}^{M} s_{ij}^2\right] \tag{2.4.18}$$

is a maximum for $i = k$.

Using Eqs. (2.4.18) and (2.3.9), we can now draw a block diagram of the optimum receiver as shown in Fig. 2.13. In the figure we have defined

$$b_i \triangleq \frac{1}{2}\left[\mathcal{N}_0 \ln P(m_i) - \sum_{j=1}^{M} s_{ij}^2\right] \tag{2.4.19}$$

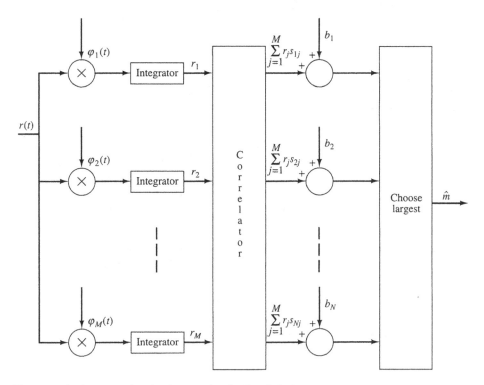

Fig. 2.13 Optimum receiver implementation for Eq. (2.4.18)

for $i = 1, 2, ..., N$. The reader should note that there are M inputs to the correlator and N outputs. The last block simply compares the N quantities and chooses $\hat{m}$ to be the m_k corresponding to the largest, as indicated by Eq. (2.4.18).

When the orthonormal basis functions $\{\varphi_j(t), j = 1, 2, ..., M\}$ are of finite duration, say, limited to be nonzero only over the interval $0 \le t \le T$, the multiplications in the implementation of Fig. 2.13 can be avoided by using matched filters to generate the components of r. To see this, consider a filter with impulse response $h_j(t), j = 1, 2, ..., M$, excited by the random process $r(t)$.

The output is then represented by the convolution

$$y_j(t) = \int_{-\infty}^{\infty} r(\lambda) h_j(t - \lambda) d\lambda. \tag{2.4.20}$$

Choosing

$$h_j(t) = \varphi_j(T - t), \quad j = 1, 2, ..., M, \tag{2.4.21}$$

we obtain from Eq. (2.4.20) that

$$y_j(t) = \int_{-\infty}^{\infty} r(\lambda)\varphi_j(T - t + \lambda)d\lambda. \tag{2.4.22}$$

If the output of the filter is sampled at $t = T$, then

$$y_j(T) = \int_{-\infty}^{\infty} r(\lambda)\varphi_j(\lambda)d\lambda, \tag{2.4.23}$$

and we see by comparison with Eq. (2.3.9) that we have produced the components of r, since $y_j(T) = r_j, j = 1, 2, \ldots, M$. We thus have an alternative method for generating the $\{r_j\}$ without using multipliers as in Fig. 2.13. The remainder of Fig. 2.13 is unchanged.

A filter with the impulse response in Eq. (2.4.21) is said to be *matched* to the signal $\varphi_j(t)$ and hence is called a *matched filter* for $\varphi_j(t)$. The matched filter can also be shown to be the filter that maximizes the filter output signal-to-noise ratio when the input is signal plus noise. Since this derivation is available in numerous texts, we do not present it here (see Problem 2.30).

The optimal receiver can also be implemented in terms of the waveforms $\{s_i(t), i = 1, 2, \ldots, N\}$ rather than the orthonormal basis functions $\{\varphi_j(t), j = 1, 2, \ldots, M\}$. To see this, consider the first term in the optimal receiver computation in Eq. (2.4.18) and write

$$\sum_{j=1}^{M} r_j s_{ij} = \sum_{j=1}^{M} s_{ij} \int_{-\infty}^{\infty} r(t)\varphi_j(t)\, dt$$

$$= \int_{-\infty}^{\infty} r(t) \sum_{j=1}^{M} s_{ij}\varphi_j(t)\, dt$$

$$= \int_{-\infty}^{\infty} r(t)s_i(t)\, dt \tag{2.4.24}$$

where we have employed Eqs. (2.3.9) and (2.3.3), respectively. Thus Eq. (2.4.24) implies that the optimal receiver can be implemented by performing N correlations of the received signal $r(t)$ with each of the possible transmitted waveforms. Following an argument similar to that in Eqs. (2.4.20)–(2.4.23), the correlation calculation in Eq. (2.4.24) can be shown to be equivalent to a matched filter implementation if the $s_i(t)$ are constrained to have a time duration T. The reader should draw block diagrams for correlation and matched filter implementations of the optimal receiver in terms of the $\{s_i(t), i = 1, 2, \ldots, N\}$ (see Problem 2.31).

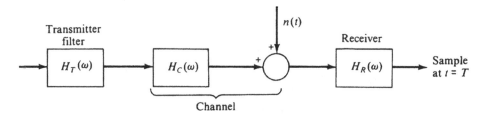

Fig. 2.14 Transfer function representation of the overall communication system

Let us now turn to a consideration of how to use the optimum receiver structures derived in this section, and matched filtering in particular, in conjunction with the pulse shaping requirements in Chap. 1. We consider various degrees of generality and present only the results, leaving the derivations to the literature. To begin, we establish some notation by referring to Fig. 2.14, where $H_T(\omega)$, $H_C(\omega)$, and $H_R(\omega)$ are the transfer functions of the transmitter filter, the channel, and the receiver filter, respectively. Suppose now that we wish to utilize a particular pulse shaping characteristic, say $P(\omega)$, and we are sending data over the AWGN channel described by Eqs. (2.3.1) and (2.3.2). In this case, $H_C(\omega) = 1$ for all ω, and the optimum transmitting and receiving filters, which minimize the probability of error and maximize the output peak signal-to-noise ratio, are $H_T(\omega) = \sqrt{P(\omega)}$ and $H_R(\omega) = \sqrt{P(\omega)}$, for $P(\omega)$ purely real. Thus, to achieve a pulse shaping $P(\omega)$ at the receiver output and to minimize the symbol error probability, the pulse shaping is evenly split, in a geometric mean sense, between the transmitter and receiver. For simplicity, we are ignoring the pure time delay of T seconds in a matched filter [see Eq. (2.4.21)], and thus the sampling in Fig. 2.14 occurs at $t = T = 0$.

The next level of generality is to consider a general $H_C(\omega)$ with additive white Gaussian noise, but only optimize for the case where a single, isolated pulse is transmitted. For $H_T(\omega)$ and $H_C(\omega)$ given, the resulting optimum receiver has the transfer function $H_R(\omega) = H_T^*(\omega)H_C^*(\omega)$. By allowing the possibility of colored noise, so that $S_n(\omega) \neq$ constant but with all other conditions the same, it can be shown that $H_R(\omega) = \left[H_T^*(\omega)H_C^*(\omega)\right]/S_n(\omega)$, where $S_n(\omega) \neq 0$ for any ω of interest. A main point to be noted here is that the receiving filter is matched not only to the desired pulse shape, but also to the shaping contributed by the channel. A block diagram of this system is shown in Fig. 2.15.

Of course, the treatment for a single isolated pulse is far from reality, since we are virtually always interested in transmitting long sequences of symbols. For $H_C(\omega) \neq$ constant, the pulses overlap and we have intersymbol interference. The optimum symbol-by-symbol receiver for the case of $H_T(\omega)$ and $H_C(\omega)$ given, additive white Gaussian noise, and a sequence of pulses is $H_R(\omega) = H_T^*(\omega)H_C^*(\omega)H_{eq}(\omega)$, where $H_{eq}(\omega)$ is the transfer function of a transversal filter equalizer. The communication system thus has the form shown in Fig. 2.16. Computing the tap gains of $H_{eq}(\omega)$ to minimize the error probability is not

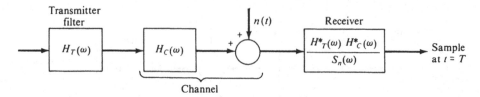

Fig. 2.15 Transfer function representation of an optimal symbol-by-symbol receiver for colored noise and single pulse transmission

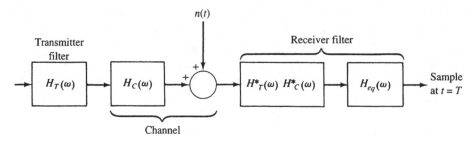

Fig. 2.16 Optimum symbol-by-symbol receiver for white noise and a sequence of transmitted pulses

a simple exercise, and hence it is common to adjust the $H_{eq}(\omega)$ coefficients to minimize peak distortion or mean-squared error as in Chap. 1.

Finally, whenever intersymbol interference is present at the receiver, whether it is due to the channel or a result of inserting controlled intersymbol interference, as in partial response signaling, symbol-by-symbol detection can be improved. Specifically, the optimum receiver in such a case examines a long sequence of symbols and makes a decision concerning the entire transmitted sequence rather than just one symbol at a time. Note that if each symbol has L possible levels and the length of the sequence is K symbols, there are L^K possible (distinct) sequences to be considered by the optimal receiver. Clearly, such receivers can be exceptionally complicated, and perhaps surprisingly, they are becoming common in off-the-shelf communications equipment. The algorithm underlying these receivers is developed in Sect. 2.7, and several systems that employ receivers based on this concept of optimum sequence detection are treated briefly in Chap. 4. For derivations and much relevant discussion, see Lucky et al. (1968) and Proakis (1989).

Up to this point in the chapter, our development has emphasized coherent demodulators, although this fact has not been stated explicitly. In retrospect, the limitation to coherent receivers is obvious, since we implicitly assume that we have available components of the received vector $\mathbf{r} = \alpha = [\alpha_1 \ \alpha_2 \ \cdots \ \alpha_M]$. This is possible only if we have coherent references for the orthonormal basis functions $\{\varphi_j(t), j = 1, 2, \ldots, M\}$ at the receiver. Now certainly the basis functions used at the transmitter are known to the receiver; however, for AMDSB-SC we know that for coherent demodulation the reference signals (LOs) must include the deterministic effects of the channel. That is, the coherent references at the receiver must be tuned to the $\{\varphi_j(t)\}$ *after* they have been transmitted over the channel. Of course, because of this requirement, coherent demodulators can be complicated, and there are many applications where simpler noncoherent demodulators are appropriate. Thus we now turn our attention to the design of optimum receivers subject to the constraint that noncoherent demodulation is to be used.

If **r** is considered to be the output of the noncoherent demodulator rather than the receiver input, the optimum signal processing required after the demodulator is still of the form given by Eq. (2.4.5). To avoid confusion, we denote the output of the noncoherent demodulator (which is a scalar) by z, so that the constrained optimum receiver sets $\hat{m} = m_k$ if and only if for all $i \neq k$,

$$f_{Z|s_k}(z|s_k) P[m_k] > f_{Z|s_i}(z|s_i) P[m_i]. \tag{2.4.25}$$

Rather than develop this result in its full generality, we now pursue the two important special cases of amplitude shift keying (ASK) and frequency shift keying (FSK). As the reader will see, the geometric interpretation so useful in the coherent receiver development is not as apparent or as important here.

A block diagram for the noncoherent ASK receiver is shown in Fig. 2.17. For ASK we have either

$$r(t) = n(t) \text{ for message } m_1, \tag{2.4.26}$$

or

$$r(t) = A_c \cos[\omega_c t + \psi] + n(t) \tag{2.4.27}$$

for message m_2 and ψ unknown. For ease of development and without loss of generality, we let $\psi = 0$ in the following. Correspondingly, the bandpass filter output is either

$$y(t) = n_o(t) \tag{2.4.28}$$

or

$$y(t) = A_c \cos \omega_c t + n_o(t), \tag{2.4.29}$$

where $n_o(t)$ is narrowband Gaussian noise as given by Eq. (2.2.7). When $m_1(s_1)$ is sent, the random variable z has a Rayleigh distribution as given by Eq. (2.2.13) or

$$f_{Z|s_1}(z|s_1) = \frac{z}{N_0/2} e^{-z^2/N_0}. \tag{2.4.30}$$

The conditional pdf of z given $m_2(s_2)$ requires a little more work. In this case

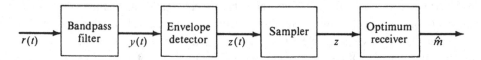

Fig. 2.17 Receiver for noncoherent ASK

$$z(t) = \sqrt{[A_c + n_i(t)]^2 + n_q^2(t)}$$

$$\triangleq \sqrt{x^2(t) + n_q^2(t)}, \tag{2.4.31}$$

so to find the pdf of z, we can find the joint pdf of $x = A + n_i$ and n_q and then perform a transformation of variables. Since n_i and n_q are both Gaussian and statistically independent, the desired joint pdf is

$$f[x, n_q] = \frac{1}{\pi \mathcal{N}_0} e^{-\left[(x - A_c)^2 + n_q^2\right]/\mathcal{N}_0}. \tag{2.4.32}$$

Defining

$$\theta \triangleq \tan^{-1} \frac{n_q}{x}, \tag{2.4.33}$$

we find that since $z = \sqrt{x^2 + n_q^2}$, $x = z \cos \theta$ and $n_q = z \sin \theta$, we can use a transformation of variables to write the joint pdf in Eq. (2.4.32) as

$$f(z, \theta) = \frac{z}{\pi \mathcal{N}_0} e^{-\left[z^2 - 2A_c z \cos \theta + A_c^2\right]/\mathcal{N}_0} \tag{2.4.34}$$

for $z \geq 0$ and $-\pi < \theta \leq \pi$. To obtain the desired conditional pdf for z, we must integrate $f(z, \theta)$ over θ. Thus

$$f_{Z|s_2}(z|s_2) = \int_{-\pi}^{\pi} f(z, \theta) d\theta$$

$$= \frac{z}{\pi \mathcal{N}_0} e^{-(z^2 + A_c^2)/\mathcal{N}_0} \int_{-\pi}^{\pi} e^{2A_c z \cos \theta / \mathcal{N}_0} d\theta$$

$$= \frac{2z}{\mathcal{N}_0} e^{-(z^2 + A_c^2)/\mathcal{N}_0} I_0 \left(\frac{2A_c z}{\mathcal{N}_0} \right), \tag{2.4.35}$$

where $z \geq 0$ and $I_0(\cdot)$ is the zeroth-order modified Bessel function of the first kind. The form of the pdf on the right side of Eq. (2.4.35) is a special one called the Rician pdf.

Using Eqs. (2.4.30) and (2.4.35) in Eq. (2.4.25) with $P(m_1) = P(m_2) = \frac{1}{2}$ yields the optimum receiver for equally likely messages and noncoherent demodulation. The

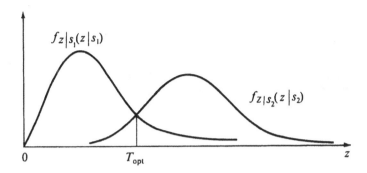

Fig. 2.18 Conditional pdfs and optimum threshold for noncoherent ASK

two pdfs and the decision rule are illustrated in Fig. 2.18, where T_{opt} is the optimum
threshold. If $z > T_{opt}$, $\hat{m} = m_2$, and if $z < T_{opt}$, $\hat{m} = m_2$. The exact determination of T_{opt}
is nontrivial. When the error probability is calculated in the next section, an approximate
value for T_{opt}, valid in the high signal-to-noise ratio case, is used.

A block diagram of a receiver for noncoherent FSK is shown in Fig. 2.19. The
transmitted signals for FSK are

$$s_1(t) = A_c \cos \omega_1 t, \quad 0 \leq t \leq T, \tag{2.4.36a}$$

for m_1 and

$$s_2(t) = A_c \cos \omega_2 t, \quad 0 \leq t \leq T, \tag{2.4.36b}$$

for m_2, where ω_1 and ω_2 are suitably spaced to minimize spectral overlap. The receiver
for noncoherent FSK is just two noncoherent ASK demodulators in parallel followed by
the sampler and "optimum" receiver. Hence, if m_1 is sent,

$$z_1(t) = A_c \cos \omega_1 t + n_o(t) \tag{2.4.37}$$

and

$$z_2(t) = n_o(t), \tag{2.4.38}$$

while if m_2 is sent,

$$z_1(t) = n_o(t) \tag{2.4.39}$$

and

$$z_2(t) = A_c \cos \omega_2 t + n_o(t). \tag{2.4.40}$$

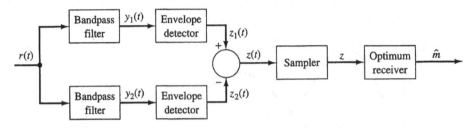

Fig. 2.19 Receiver for noncoherent FSK

In Eqs. (2.4.37)–(2.4.40), $n_o(t)$ is narrowband Gaussian noise. Following the development for noncoherent ASK, the conditional pdfs of $z_1(t)$ and $z_2(t)$ after sampling when m_1 is sent are

$$f_{Z_1|s_1}(z_1|s_1) = \frac{2z_1}{\mathcal{N}_0}e^{-(z_1^2+A_c^2)/\mathcal{N}_0}I_0\left(\frac{2A_cz_1}{\mathcal{N}_0}\right) \tag{2.4.41}$$

for $z_1 > 0$, and

$$f_{Z_2|s_1}(z_2|s_1) = \frac{2z_2}{\mathcal{N}_0}e^{-z_2^2/\mathcal{N}_0} \tag{2.4.42}$$

for $z_2 > 0$. The threshold for $z = z_1 - z_2$ is $T_{\text{opt}} = 0$, so we decide $\hat{m} = m_1$ if $z > 0$ or $z_1 > z_2$. We can write a similar relation for when m_2 is sent. Using these results, the probability of error is easy to calculate.

We conclude this section by noting that we have specified the optimum (minimum $P[\mathcal{E}]$) decision rule for transmitting N messages over an AWGN channel and have shown how these decision rules might be implemented. We now turn our attention to error probability calculations for performance evaluations and to some aspects of signal design.

2.5 Error Probability and Signal Design

To find expressions for the error probability of an optimum receiver, we consider M-dimensional space and note that Eq. (2.4.11) partitions this space into N regions (decision regions) each one of which is associated with one and only one of the transmitted signal vectors, s_i, $i = 1, 2, ..., N$. Due to the correspondence $s_i \Leftrightarrow m_i$, we see that each of the N regions is also associated with one and only one of the transmitted messages. Denoting the decision region for the ith signal (and message) by $\mathcal{R}_i$, we see that for a received vector $\mathbf{r} = \alpha$, we decide $\hat{m} = m_k$ if $\mathbf{r} = \alpha \in \mathcal{R}_k$, that is, if the received $\mathbf{r}$ vector falls in the region of M-dimensional space associated with m_k.

Therefore, we make an error if $\mathbf{r} = \alpha \notin \mathcal{R}_k$ when s_k is transmitted. Denoting the probability of error given that s_k, and hence m_k, is sent by $P[\mathcal{E}|m_k]$, we can write

$$P[\mathcal{E}|m_k] = P[\mathbf{r} = \alpha \notin \mathcal{R}_k|m_k]$$
$$= 1 - P[\mathbf{r} = \alpha \in \mathcal{R}_k|m_k]$$
$$= 1 - P[\mathcal{C}|m_k]. \tag{2.5.1}$$

Since we are interested in the total probability of error over the entire signal set, we use the a priori message probabilities and write

$$P[\mathcal{E}] = \sum_{i=1}^{N} P[\mathcal{E}|m_i]P[m_i]$$
$$= \sum_{i=1}^{N} P[\mathbf{r} = \alpha \notin \mathcal{R}_i|m_i]P[m_i]. \tag{2.5.2}$$

It is often more convenient to calculate $P[\mathcal{E}]$ by using

$$P[\mathcal{E}] = 1 - P[\mathcal{C}]$$
$$= 1 - \sum_{i=1}^{N} P[\mathcal{C}|m_i]P[m_i]$$
$$= 1 - \sum_{i=1}^{N} P[\mathbf{r} = \alpha \in \mathcal{R}_i|m_i]P[m_i]. \tag{2.5.3}$$

Thus, once we calculate the conditional error probabilities or the conditional probabilities of a correct decision, we can find the total error probability from Eq. (2.5.2) or (2.5.3).

The problem is considerably simplified by the fact that we often work with a one- or two-dimensional signal space and that we are assuming an AWGN channel. Rather than continue to pursue this topic in generality, we consider a series of physically important illustrative examples.

Example 2.5.1 We reexamine the binary PSK signal set with equally likely messages over an AWGN channel previously demonstrated in Example 2.4.2 to have the optimum decision regions shown in Fig. 2.10. From this figure we see that $\mathcal{R}_1 = \{r = \alpha > 0\}$ and $\mathcal{R}_2 = \{r = \alpha < 0\}$. Thus, since $P[m_1] = P[m_2] = \frac{1}{2}$, we have from Eq. (2.5.2) that

$$P[\mathcal{E}] = \frac{1}{2}\{P[r < 0|m_1] + P[r > 0|m_2]\}. \tag{2.5.4}$$

Assuming that the noise is zero mean with variance σ^2 (this implies that $\mathcal{N}_0/2 = \sigma^2$), we can write

$$P[r < 0|m_1] = \frac{1}{\sqrt{2\pi}\sigma} \int_{-\infty}^{0} e^{-(r-A_c)^2/2\sigma^2} dr. \tag{2.5.5}$$

To continue the problem, we must manipulate the integral in Eq. (2.5.5) into the form of the Gaussian error function. Letting $\lambda = -(r - A_c)/\sigma$ produces

$$P[r < 0|m_1] = \frac{1}{\sqrt{2\pi}} \int_{A_c/\sigma}^{\infty} e^{-\lambda^2/2} d\lambda. \tag{2.5.6}$$

Since $\int_0^\infty (1/\sqrt{2\pi})e^{-y^2/2}dy = \frac{1}{2}$, we can rewrite Eq. (2.5.6) as

$$P[r < 0|m_1] = \frac{1}{2} - \frac{1}{\sqrt{2\pi}} \int_0^{A_c/\sigma} e^{-\lambda^2/2} d\lambda$$

$$= \frac{1}{2} - \mathrm{erf}\left[\frac{A_c}{\sigma}\right]. \tag{2.5.7}$$

We leave it to the reader to show that

$$P[r > 0|m_2] = \frac{1}{2} - \mathrm{erf}\left[\frac{A_c}{\sigma}\right], \tag{2.5.8}$$

so that using Eq. (2.5.4) we obtain

$$P[\mathcal{E}] = \frac{1}{2} - \mathrm{erf}\left[\frac{A_c}{\sigma}\right]. \tag{2.5.9}$$

Example 2.5.2 As a second example, we consider the equally likely binary signal set shown in Fig. 2.20. This signal set could represent using the time-domain signals $s_1(t) = A_c\sqrt{2/T}\cos\omega_c t$ and $s_2(t) = A_c\sqrt{2/T}\sin\omega_c t, 0 \le t \le T$, for a 1 and 0, among numerous other possibilities. In any event, our goal is to calculate $P[\mathcal{E}]$ for this signal set when it is used over an AWGN channel with zero mean and variance σ^2.

To begin, we note from Eq. (2.4.12) that for equally likely signals, the decision rule depends only on the distance between the two signals and not in any way on their orientation in the space. When the signals are not equally likely, it is evident from Eqs. (2.4.10) and (2.4.11) that we can similarly conclude that it is only the distance between the signals plus a bias term which are of importance and neither depends on where the signals are in space—as long as the distance is preserved.

Fig. 2.20 Orthogonal signal
set for Example 2.5.2

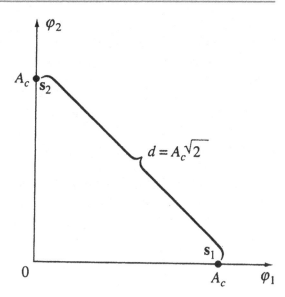

We thus conclude that we need only compute the distance between s_1 and s_2 in Fig. 2.20 and translate this into an equivalent one-dimensional problem as shown in Fig. 2.21. The signal constellations in Figs. 2.10 and 2.21 are identical except for a scale factor of $1/\sqrt{2}$. Hence we use the result from Example 2.5.1 to obtain

$$P[\mathcal{E}] = \frac{1}{2} - \mathrm{erf}\left[\frac{A_c}{\sigma\sqrt{2}}\right]. \qquad (2.5.10)$$

Comparing Eqs. (2.5.9) and (2.5.10), we see that since $\mathrm{erf}(x)$ is an increasing function of x, equal amplitude binary antipodal signals outperform (have a smaller $P[\mathcal{E}]$) binary orthogonal signals.

In solving Example 2.5.2, we argued that we could rotate and translate the signal set and the decision regions equally without affecting $P[\mathcal{E}]$. This result remains true for more complicated signal constellations and decision regions, and it can be extremely useful in simplifying calculations. In Example 2.5.2 we were able to use this fact to change a two-dimensional problem into a one-dimensional problem. That rotations and translations of the signal set and the corresponding decision regions do not affect $P[\mathcal{E}]$ follows from the decision rules in Eqs. (2.4.10)–(2.4.12), as claimed in the example. In turn, the decision rules were derived based on the assumption of additive white Gaussian noise that is

Fig. 2.21 $P[\mathcal{E}]$ equivalent
signal set for Example 2.5.2

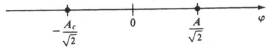

independent of the transmitted signal. Thus the noise is independent of the location of the signal in signal space, and since the Gaussian pdf is spherically symmetric, the rotation of the decision regions does not change $P[\mathcal{E}]$. This invariance of $P[\mathcal{E}]$ to rotations and translations for an AWGN channel is used often in communications.

It should be noted that while translations of signal constellations do not affect $P[\mathcal{E}]$, they can affect both the peak and average energies of the signal set, which are of physical significance. However, for $P[\mathcal{E}]$ analyses we can consider rotated and translated signal sets in order to simplify our calculations.

Example 2.5.3 We wish to calculate $P[\mathcal{E}]$ for the signal set in Fig. 2.1 when used for equally likely messages over an AWGN channel with zero mean and variance σ^2. This signal set could srepresent four-level QAM or four-level PSK, among others. Since the signals (messages) are equally likely, we can write

$$P[\mathcal{E}] = \frac{1}{4}\sum_{i=1}^{4} P[\mathcal{E}|m_i]. \qquad (2.5.11)$$

Thus we need to compute the conditional error probabilities, $P[\mathcal{E}|m_i]$. Starting with m_1, we have

$$\begin{aligned} P[\mathcal{E}|m_1] &= P[\mathbf{r} \notin \mathcal{R}_1|m_1] \\ &= 1 - P[\mathbf{r} \in \mathcal{R}_1|m_1]. \end{aligned} \qquad (2.5.12)$$

The region $\mathcal{R}_1$ where we accept m_1 is given by the first quadrant, which is simply the set $r_1 > 0$ and $r_2 > 0$, where $\mathbf{r} = [r_1 \ r_2]$. Therefore,

$$P[\mathbf{r} \in \mathcal{R}_1|m_1] = P[r_1 > 0, r_2 > 0|m_1]. \qquad (2.5.13)$$

Since the components of the noise vector are statistically independent of each other, we can rewrite Eq. (2.5.13) as

$$P[\mathbf{r} \in \mathcal{R}_1|m_1] = P[r_1 > 0|m_1]P[r_2 > 0|m_1]. \qquad (2.5.14)$$

Now, given that m_1 is transmitted, $r_1 = s_{11} + n_1$, where n_1 is zero mean, Gaussian with variance σ^2. As a result, r_1 is Gaussian with mean s_{11} and variance σ^2, so

$$P[r_1 > 0|m_1] = \int_{0}^{\infty} \frac{1}{\sqrt{2\pi}\sigma} e^{-(r_1-s_{11})^2/2\sigma^2} dr_1, \qquad (2.5.15)$$

which becomes upon using the definition of the Gaussian error function in Eq. (A.6.11) that

$$P[r_1 > 0|m_1] = \frac{1}{2} + \text{erf}\left[\frac{s_{11}}{\sigma}\right]$$

$$= \frac{1}{2} + \text{erf}\left[\frac{A_c}{\sigma}\right]. \tag{2.5.16}$$

The last equality follows from Fig. 2.1, since $s_{11} = A_c$. The integral required to compute $P[r_2 > 0|m_1]$ is identical to the one in Eq. (2.5.15), so we have that

$$P[\mathbf{r} \in \mathcal{R}_1|m_1] = \left\{\frac{1}{2} + \text{erf}\left[\frac{A_c}{\sigma}\right]\right\}^2,$$

so

$$P[\mathcal{E}|m_1] = 1 - \left\{\frac{1}{2} + \text{erf}\left[\frac{A_c}{\sigma}\right]\right\}^2$$

$$= \frac{3}{4} - \text{erf}\left[\frac{A_c}{\sigma}\right] - \left\{\text{erf}\left[\frac{A_c}{\sigma}\right]\right\}^2. \tag{2.5.17}$$

Although we can proceed to calculate the other conditional error probabilities directly just as for $P[\mathcal{E}|m_1]$, we take an alternative approach of noting the symmetry of the signal set in Fig. 2.1. Furthermore, we have assumed that the noise is Gaussian, and hence circularly symmetric, and that the noise is independent of the particular signal being transmitted. Based on these symmetry arguments, we thus claim that $P[\mathcal{E}|m_2] = P[\mathcal{E}|m_3] = P[\mathcal{E}|m_4] = P[\mathcal{E}|m_1]$ in Eq. (2.5.17). Therefore, from Eq. (2.5.11), the total probability of error is

$$P[\mathcal{E}] = P[\mathcal{E}|m_1] = \frac{3}{4} - \text{erf}\left[\frac{A_c}{\sigma}\right] - \left\{\text{erf}\left[\frac{A_c}{\sigma}\right]\right\}^2. \tag{2.5.18}$$

We now turn our attention to calculating the probability of error for noncoherent ASK and noncoherent FSK. To simplify the development for ASK, we limit consideration to the high signal-to-noise ratio case, where $A_c/\sqrt{\mathcal{N}_0} \gg 1$. In this situation, the conditional pdf given s_2 in Eq. (2.4.34) is approximately Gaussian and is given by

$$f_{Z|s_2}(z|s_2) \cong \frac{1}{\sqrt{\pi\mathcal{N}_0}} e^{-(z-A_c)^2/\mathcal{N}_0} \tag{2.5.19}$$

and the optimum threshold, T_{opt}, is about $A_c/2$. The probability of error for noncoherent ASK is thus

$$P[\mathcal{E}] = \frac{1}{2}P[\mathcal{E}|s_1] + \frac{1}{2}P[\mathcal{E}|s_2], \tag{2.5.20}$$

where by using Eq. (2.4.30),

$$P[\mathcal{E}|s_1] = \int_{A_c/2}^{\infty} \frac{2z}{\mathcal{N}_0} e^{-z^2/\mathcal{N}_0} dz = e^{-A_c^2/4\mathcal{N}_0} \tag{2.5.21}$$

and using Eq. (2.5.19),

$$P[\mathcal{E}|s_2] \cong \int_{-\infty}^{A_c/2} \frac{1}{\sqrt{\pi \mathcal{N}_0}} e^{-(z-A_c)^2/\mathcal{N}_0} dz$$

$$\cong \sqrt{\frac{\mathcal{N}_0}{\pi}} \cdot \frac{1}{A_c} e^{-A_c^2/4\mathcal{N}_0}. \tag{2.5.22}$$

To obtain the result in Eq. (2.5.22), we have evaluated the integral in terms of the error function and employed the approximation

$$1 - \operatorname{erf} x \cong \frac{\sqrt{2}}{x\sqrt{\pi}} e^{-x^2/2} \quad \text{for } x \gg 1.$$

Substituting Eqs. (2.5.21) and (2.5.22) into Eq. (2.5.20) yields

$$P[\mathcal{E}] = \frac{1}{2} e^{-A_c^2/4\mathcal{N}_0} + \frac{1}{2A_c} \sqrt{\frac{\mathcal{N}_0}{\pi}} e^{-A_c^2/4\mathcal{N}_0}$$

$$= \left[1 + \frac{1}{A_c} \sqrt{\frac{\mathcal{N}_0}{\pi}} \right] \cdot \frac{1}{2} e^{-A_c^2/4\mathcal{N}_0}. \tag{2.5.23}$$

Once again invoking the high signal-to-noise ratio assumption, the second term in brackets is small, so

$$P[\mathcal{E}] \cong \frac{1}{2} e^{-A_c^2/4\mathcal{N}_0}. \tag{2.5.24}$$

Although Eq. (2.5.24) is only valid for $A_c/\sqrt{\mathcal{N}_0} \gg 1$, it is often used as the $P[\mathcal{E}]$ expression for noncoherent ASK, and we will follow suit here.

For noncoherent FSK, we need not invoke such approximations, since the threshold is just the variable in the parallel channel. Referring to Fig. 2.19 and Eqs. (2.4.41) and (2.4.42), we see that an error is made when m_1 is sent if $z_1 < z_2$ and if m_2 is sent when $z_1 > z_2$. Thus

$$P[\mathcal{E}|s_1] = \int_0^\infty f_{Z_1|s_1}(z_1|s_1) \left[\int_{z_1}^\infty f_{Z_2|s_1}(z_2|s_1) dz_2 \right] dz_1, \tag{2.5.25}$$

which upon substituting for $f_{Z_2|s_1}(z_2|s_1)$ and integrating yields

$$P[\mathcal{E}|s_1] = e^{-A_c^2/\mathcal{N}_0} \int_0^\infty \frac{2z_1}{\mathcal{N}_0} e^{-z_1^2/(\mathcal{N}_0/2)} I_0\left(\frac{2A_c z_1}{\mathcal{N}_0}\right) dz_1. \tag{2.5.26}$$

This integral is available in tables, so Eq. (2.5.26) becomes

$$P[\mathcal{E}|s_1] = \frac{1}{2}e^{-A_c^2/2\mathcal{N}_0}. \tag{2.5.27}$$

By symmetry, $P[\mathcal{E}|s_2] = P[\mathcal{E}|s_1]$, so since the messages are equally likely

$$P[\mathcal{E}] = \frac{1}{2}e^{-A_c^2/2\mathcal{N}_0}. \tag{2.5.28}$$

This and other $P[\mathcal{E}]$ results are compared in Sect. 2.6.

Given a certain transmitted signal set and a particular channel, we know how to design an optimum receiver that minimizes $P[\mathcal{E}]$ and indeed this has been the main thrust of the chapter thus far. The problem can be turned around, however, to obtain an interesting problem in what is called *signal design*. Specifically, given a particular channel and the

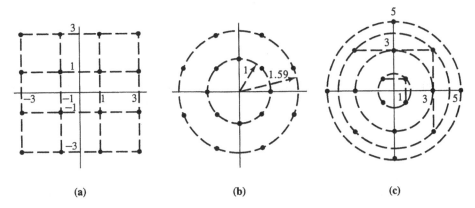

| (a) | (b) | (c) |

Fig. 2.22 Common 16-point signal constellations: **a** four-level **QAM;** **b** modified 8–8 AM/PM; **c** circular (4, 90°) constellation

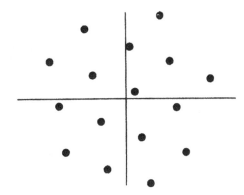

Fig. 2.23 Optimum 16-point signal constellation for an average energy constraint. From G. J. Foschini, R. D. Gitlin, and S. B. Weinstein, "Optimization of Two-Dimensional Signal Constellations in the Presence of Gaussian Noise," *IEEE Trans. Commun.,* © 1974 IEEE

minimum $P[\mathcal{E}]$ receiver design philosophy, how do we choose the transmitted signal set
or signal constellation to minimize $P[\mathcal{E}]$? Here we limit consideration to the coherent
receiver case, since this allows us to take maximum advantage of the geometric interpretation and to obtain significant practical results. The problem can be posed with either an
average energy or peak energy constraint on the transmitted signals, but in either case, the
optimization problem is not a simple one, since conditional pdfs must be integrated over
peculiarly shaped decision regions and the sum of these integrals must be minimized. In
the high-SNR case, only adjacent decision regions need to be considered, but the problem
remains difficult. On the other hand, the problem is relatively easy to state verbally as one
of maximizing the spacing between adjacent points in signal space subject to the appropriate energy constraint. Three well-known 16-point signal constellations not obtained
via this optimization procedure are shown in Fig. 2.22 for benchmark purposes. The optimum 16-point constellation under an average energy constraint for an AWGN channel
is shown in Fig. 2.23 (Foschini et al. 1974). Note the lack of symmetry in comparison
with the signal constellations in Fig. 2.22. The optimum signal constellation subject to a
peak energy constraint for an AWGN channel is presented in Fig. 2.24 (Kernighan and
Lin 1973). Notice that this constellation is less "random looking" than the constellation
in Fig. 2.23, but not nearly as symmetric as those in Fig. 2.22. If the channel causes
random carrier phase jitter in addition to injecting AWGN, the optimum signal set subject
to a peak energy constraint is that shown in Fig. 2.25 (Kernighan and Lin 1973). When
phase jitter is included, points farther from the origin suffer increased phase distortion,
and hence one of the outer signal points in Fig. 2.24 is moved to the origin in Fig. 2.25.
Calculations pertaining to these signal sets and signal design in particular are left to the
references.

Fig. 2.24 Optimum 16-point
signal constellation for a peak
energy constraint (peak SNR =
27 dB). From B. W. Kernighan
and S. Lin, "Heuristic Solution
of a Signal Design
Optimization Problem," Proc.
7th Annual Princeton
Conference on Information
Science and Systems, March
1973

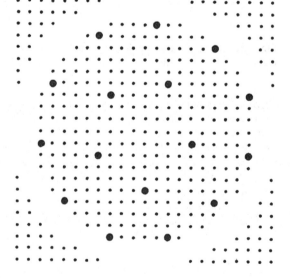

Fig. 2.25 Optimum 16-point signal constellation for a peak energy constraint (SNR = 27 dB, RMS phase jitter = 3°). From B. W. Kernighan and S. Lin, "Heuristic Solution of a Signal Design Optimization Problem," Proc. 7th Annual Princeton Conference on Information Science and Systems, March 1973

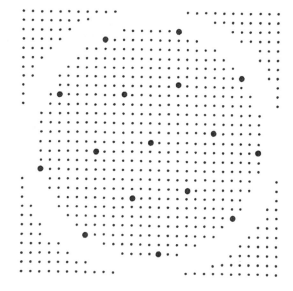

2.6   System Performance Comparisons

The first question that arises when an engineer is designing a digital communication system is: Which signaling scheme is best? The answer to this question involves many considerations such as probability of error, energy constraints, complexity, bandwidth, channel nonlinearities, and others, and these details are usually known only for very specific communications problems. Here we briefly provide some comparisons based primarily on $P[\mathcal{E}]$, with some discussion concerning energy constraints, bandwidth, and complexity. The signaling methods examined are coherent and noncoherent ASK, coherent and noncoherent FSK, coherent BPSK, and differential PSK(DPSK). We first focus on the bit error probabilities of these schemes, which are usually plotted versus the ratio of energy per bit to one-sided noise power spectral density, $E_b/\mathcal{N}_0$.

We begin by writing the previously derived $P[\mathcal{E}]$ expressions in terms of E_b and $\mathcal{N}_0$, which we accomplish by noting that $A_c = \sqrt{E_b}$, since we are transmitting one bit per dimension and the basis functions are orthonormal, and that $\sigma^2 = \mathcal{N}_0/2$. Using these definitions, we have from Eq. (2.5.9) that for coherent BPSK,

$$P[\mathcal{E}] = \frac{1}{2} - \mathrm{erf}\left[\sqrt{\frac{2E_b}{\mathcal{N}_0}}\right]. \qquad (2.6.1)$$

For coherent FSK where the signals are sufficiently separated in frequency to be considered orthogonal, Eq. (2.5.10) yields

$$P[\mathcal{E}] = \frac{1}{2} - \text{erf}\left[\sqrt{\frac{E_b}{\mathcal{N}_0}}\right].\tag{2.6.2}$$

It is straightforward to show that for coherent ASK with a peak power constraint that

$$P[\mathcal{E}] = \frac{1}{2} - \text{erf}\left[\sqrt{\frac{E_b}{2\mathcal{N}_0}}\right].\tag{2.6.3}$$

while for an average power constraint coherent ASK has a $P[\mathcal{E}]$ given by Eq. (2.6.2). For noncoherent ASK subject to a peak power constraint, we have from Eq. (2.5.24),

$$P[\mathcal{E}] \cong \frac{1}{2}e^{-E_b/4\mathcal{N}_0},\tag{2.6.4}$$

while with an average power constraint,

$$P[\mathcal{E}] \cong \frac{1}{2}e^{-E_b/2\mathcal{N}_0}.\tag{2.6.5}$$

The error probability for noncoherent FSK can be obtained from Eq. (2.5.28) as

$$P[\mathcal{E}] = \frac{1}{2}e^{-E_b/2N_0}.\tag{2.6.6}$$

Finally, we display some error probability results for a binary DPSK system. Although the derivation is somewhat involved, and hence not given here, the bit error probability for DPSK is simple and is given by

$$P[\mathcal{E}] = \frac{1}{2}e^{-E_b/\mathcal{N}_0}.\tag{2.6.7}$$

Plots of these error probabilities subject to an average energy constraint are shown in Fig. 2.26. For a peak power constraint, the coherent and noncoherent ASK curves are shifted to the right by 3 dB.

Since there are many possible definitions of bandwidth for phase- and frequency-modulated carriers and random pulse sequence messages, comparative bandwidth statements are necessarily imprecise. In general, FSK methods require a greater bandwidth than ASK, DPSK, and BPSK, which have similar bandwidth requirements. In terms of complexity, noncoherent ASK is the least complex, followed by noncoherent FSK, DPSK, BPSK, and coherent ASK and FSK. Note that for high signal-to-noise ratios ($E_b/\mathcal{N}_0$), noncoherent ASK and FSK perform within 1 dB of their coherent counterparts and hence are quite attractive as a compromise between complexity and performance. Similarly, DPSK has performance close to BPSK at high $E_b/\mathcal{N}_0$ and is simpler to implement, since a coherent reference is not required. These statements provide some indication as to why noncoherent FSK and DPSK have found widespread acceptance in many practical communication systems.

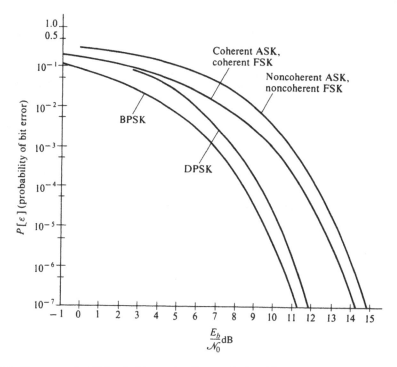

Fig. 2.26 Bit error probabilities for binary signaling methods subject to an average power constraint

2.7   Maximum Likelihood Sequence Estimation

In this section we develop receiver structures that are truly optimum in the presence of intersymbol interference and additive white Gaussian noise. As noted in Sect. 2.4, the intersymbol interference may be caused by the channel or it may have been inserted purposely at the transmitter by using partial response signaling. The underlying algorithm is the same in both cases, and the improvement in performance over optimum symbol-by-symbol detection is due to the receiver making decisions for an entire sequence of symbols rather than just on individual symbols.

We must now include a time index in our notation, so that if the ith message is input to the transmitter at time k, we denote this as $m_i(k)$, and the corresponding transmitted signal vector at time k is $\mathbf{s}_i(k)$ each transmitted symbol interferes with J other symbols, the received vector at time k for a general transmitted signal sequence $\{\mathbf{s}(k)\}$ can be written as

$$\mathbf{r}(k) = \sum_{j=0}^{J} d_j \mathbf{s}(k - j) + \mathbf{n}(k), \tag{2.7.1}$$

where the coefficients d_j, $j = 0, 1, \ldots, J$ account for the intersymbol interference and $\mathbf{n}(k)$ is the Gaussian noise vector at time k. We assume that all messages are equally likely and consider a transmitted sequence K symbols long, $K > J$. If we denote the received and transmitted sequences, respectively, as $\mathbf{R}_K = (\mathbf{r}(1), \mathbf{r}(2),\ldots, \mathbf{r}(K)\}$ and $\mathbf{S}_K = \{\mathbf{s}(1), \mathbf{s}(2),\ldots, \mathbf{s}(K)\}$, following Sect. 2.4, the optimum receiver for the sequence maximizes

$$f_{\mathbf{R}|\mathbf{S}}(\mathbf{R}_K|\mathbf{S}_K) \triangleq f_{\mathbf{R}|\mathbf{S}}(\mathbf{r}(K), \ldots, \mathbf{r}(1)|\mathbf{s}(K), \ldots, \mathbf{s}(1)). \qquad (2.7.2)$$

As before, the additive Gaussian noise is assumed to be white, so that Eq. (2.7.2) can be expressed as

$$f_{\mathbf{R}|\mathbf{S}}(\mathbf{R}_K|\mathbf{S}_K) = \prod_{k=1}^{K} f_{\mathbf{r}|\mathbf{S}_J}(\mathbf{r}(k)|\mathbf{s}(k), \ldots, \mathbf{s}(k-J)). \qquad (2.7.3)$$

The optimum sequence estimator in Eq. (2.7.3) clearly subsumes the optimum symbol-by-symbol receiver, since if there is no intersymbol interference, the conditional pdfs on the right side of Eq. (2.7.3) reduce to $f_{\mathbf{r}|\mathbf{s}}(\mathbf{r}(k)|\mathbf{s}(k))$, as in Sect. 2.4. It is common to take (natural) logarithms of both sides of Eq. (2.7.3), to yield

$$\ln f_{\mathbf{R}|\mathbf{S}}(\mathbf{R}_K|\mathbf{S}_K) = \sum_{k=1}^{K} \ln f_{\mathbf{r}|\mathbf{S}_J}(\mathbf{r}(k)|\mathbf{s}(k), \ldots, \mathbf{s}(k-J)) \qquad (2.7.4)$$

as the quantity to be maximized by the selection of the decoded sequence $\mathbf{S}_K$.

To perform the maximization, we note that $\mathbf{s}(k) = 0$ for $k \leq 0$, and write out a few terms of the summation explicitly to find

$$\ln f_{\mathbf{R}|\mathbf{S}}(\mathbf{R}_K|\mathbf{S}_K) = \sum_{k=3}^{K} \ln f_{\mathbf{r}|\mathbf{S}_J}(\mathbf{r}(k)|\mathbf{s}(k), \ldots, \mathbf{s}(k-J))$$
$$+ \ln f_{\mathbf{r}|\mathbf{S}_J}(\mathbf{r}(2)|\mathbf{s}(2), \mathbf{s}(1)) + \ln f_{\mathbf{r}|\mathbf{S}_J}(\mathbf{r}(1)|\mathbf{s}(1)), \qquad (2.7.5)$$

which we wish to maximize with respect to the selection of $\mathbf{s}(1), \ldots, \mathbf{s}(K)$. Thus we can write

$$\max_{\mathbf{S}_K} \ln f_{\mathbf{R}|\mathbf{S}}(\mathbf{R}_K|\mathbf{S}_K) = \max_{s(1)} \left\{ \ln f_{\mathbf{r}|\mathbf{S}_J}(r(1)|s(1)) + \max_{s(2)} \left\{ \ln f_{\mathbf{r}|\mathbf{S}_J}(r(2)|s(2), s(1)) + \cdots \right. \right.$$
$$\left. \left. + \max_{s(K)} \left\{ \ln f_{\mathbf{r}|\mathbf{S}_J}(r(K)|s(K), \ldots, s(K-J)) \right\} \underbrace{\cdots}_{K \text{ braces}} \right\} \right\}$$
$$= \max_{s(1)} \ln f_{\mathbf{r}|\mathbf{S}_J}(r(1)|s(1)) + \max_{s(1),s(2)} \ln f_{\mathbf{r}|\mathbf{S}_J}(r(2)|s(2), s(1)) + \cdots$$
$$+ \max_{s(1),\ldots,s(K)} \ln f_{\mathbf{r}|\mathbf{S}_J}(r(K)|s(K), \ldots, s(K-J)) \qquad (2.7.6)$$

There are N possible transmitted signal vectors $\mathbf{s}(k)$ at each time instant k, so that for a sequence of length K, there are N^K possible sequences that must be considered in the most general case. However, for our current situation where there is intersymbol interference with $J < K$ symbols, the computations required are somewhat less. To see this, we expand the last part of Eq. (2.7.6) as

$$
\max_{S_K} \ln f_{RS}(\mathbf{R}_K | \mathbf{S}_K) = \max_{s(1)} \ln f_{r|S_J}(r(1)|s(1)) + \max_{s(1),s(2)} \ln f_{r|S_J}(r(2)|s(2), s(1)) + \cdots
$$
$$
+ \max_{s(1),\dots,s(J+1)} \ln f_{r|S_J}(r(J+1)|s(J+1), \dots, s(1))
$$
$$
+ \max_{s(1),s(2),\dots,s(J+2)} \ln f_{r|S_J}(r(J+2)|s(J+2), \dots, s(2)) + \cdots
$$
$$
+ \max_{s(1),\dots,s(K)} \ln f_{r|S_J}(r(K)|s(K), \dots, s(K-J)). \tag{2.7.7}
$$

After time instant $k = J$, each term requires the computation of N^{J+1} probabilities, so that the number of probabilities needed for Eq. (2.7.7) is upper bounded by KN^{J+1}. If $J \ll K$, this upper bound is fairly tight.

Let us now consider all the terms in Eq. (2.7.7) with explicit conditioning on $\mathbf{s}(1)$. If these $J + 1$ maximizations yield the same value for $\mathbf{s}(1)$, that is the decoded value corresponding to $\mathbf{r}(1)$. If, however, these maximizations yield different values for $\mathbf{s}(1)$, the decision on $\mathbf{s}(1)$ must be deferred. At the next time instant, $k = J + 2$, we try to make a decision on $\mathbf{s}(1)$ and $\mathbf{s}(2)$. Again, if all the maximizations yield the same $\mathbf{s}(1)$ or $\mathbf{s}(1)$ and $\mathbf{s}(2)$, then $\mathbf{s}(1)$ or $\mathbf{s}(1)$ and $\mathbf{s}(2)$, respectively, can be definitely decided. If not, the decision is again deferred. Now the question becomes: How long can the decision on $\mathbf{s}(1)$ and following symbols be delayed? The answer is that since we are making decisions on sequences and each sequence overlaps the immediately preceding sequence by J symbols, the memory is essentially infinite, so the decision on $\mathbf{s}(1)$ could be delayed until the end of the sequence, time instant $k = K$. However, since K is typically very large, an upper limit is usually chosen on the delay in deciding each particular symbol, and it has been determined that if this delay is greater than or equal to $5J$, there is negligible loss in performance. If at some time instant k, all candidate sequences disagree on the symbol $\mathbf{s}(k - 5J)$, then the value of $\mathbf{s}(k - 5J)$ in the most probable sequence is picked as the decoded value.

It is important to notice that implicit in the optimization in Eq. (2.7.7) is the fact that at some time instant k, only the N best sequences through time instant $k - J$ need to remain under consideration. This is a manifestation of what is called *the principle of optimality*, which states that the optimal sequence including some value of $\mathbf{s}(k - J)$ has as a subset the optimal sequence from the beginning of the decoding process (time 0) to the candidate output $\mathbf{s}(k - J)$. The explicit statement of this concept is an aid when one is solving specific problems.

Example 2.7.1 To illustrate the use of this optimum sequence estimation algorithm, we consider the use of duobinary or class I partial response signaling to transmit binary (± 1) data (see Sect. 1.3 for details on duobinary). The received values at time instants 1 and 2 are thus

$$r(1) = s(1) + n(1) \tag{2.7.8}$$

and

$$r(2) = s(2) + s(1) + n(2). \tag{2.7.9}$$

We assume here that K is very large, and we know for duobinary that $J = 1$. The pertinent optimization is thus summarized as

$$\max_{\mathbf{S}_K} \ln f_{\mathbf{R}|\mathbf{S}}(\mathbf{R}_K|\mathbf{S}_K) = \max_{s(1)} \ln f_{\mathbf{r}|\mathbf{S}_1}(r(1)|s(1)) + \max_{s(1),s(2)} \ln f_{\mathbf{r}|\mathbf{S}_1}(r(2)|s(2), s(1))$$

$$+ \max_{s(1),s(2),s(3)} \ln f_{\mathbf{r}|\mathbf{S}_1}(r(3)|s(3), s(2)) + \cdots$$

$$+ \max_{s(1),\ldots,s(k)} \ln f_{\mathbf{r}|\mathbf{S}_1}(r(k)|s(k), s(k-1)) + \cdots$$

$$+ \max_{s(1),\ldots,s(K)} \ln f_{\mathbf{r}|\mathbf{S}_1}(r(K)|s(K), s(K-1)). \tag{2.7.10}$$

Since for any k, $s(k) = \pm 1$, we can represent all possible output sequences from time 0 to time k by a tree diagram as shown in Fig. 2.27. Noting that the noise is zero-mean Gaussian and the inputs are equally likely, we have that an individual term in Eq. (2.7.10) is given by

$$\max_{s(1),\ldots,s(k)} \ln f_{\mathbf{r}|\mathbf{S}_1}(r(k)|s(k), s(k-1)) = \max_{s(1),\ldots,s(k)} \left\{ -[r(k) - s(k) - s(k-1)]^2 \right\}.$$

$$\tag{2.7.11}$$

The first maximization to be examined is therefore

$$\max_{s(1),s(2)} \left\{ -[r(1) - s(1)]^2 - [r(2) - s(2) - s(1)]^2 \right\}, \tag{2.7.12}$$

and it has four possible values corresponding to the four candidate paths to depth (time instant) 2 in the tree of Fig. 2.27. Now, at the next time instant, the received value does not involve $s(1)$, so that we need only retain at time instant 2, $N = 2$ paths, consisting of the best path terminating with $s(2) = +1$ and the best path ending at $s(2) = -1$. Since we do not pick explicit values for the received values in this example, we have arbitrarily indicated the extension of two paths at $k = 2$ in Fig. 2.27. Note that these paths do not have the same value for $s(1)$, so this output cannot be decided as yet. At time instant 3, the maximization is

$$\max_{s(1),s(2),s(3)} \left\{ -[r(3) - s(3) - s(2)]^2 \right.$$

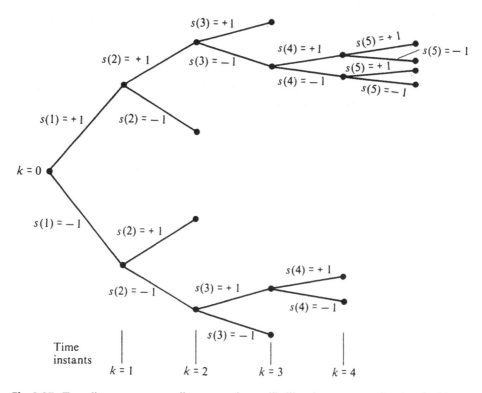

Fig. 2.27 Tree diagram corresponding to maximum likelihood sequence estimation for binary-driven duobinary

$$+ \max_{s(1),s(2)} \left\{ -[r(1) - s(1)]^2 - [r(2) - s(2) - s(1)]^2 \right\}, \tag{2.7.13}$$

so that we only need to extend the two best paths as shown in Fig. 2.27. Again, we note that the next received value at time instant 4 does not depend on $s(3)$, so we retain only the $N = 2$ best sequences leading to $s(3) = -1$ and $s(3) = +1$. A possible pruning of paths is indicated in the figure, and the two best paths still do not have the same value of $s(1)$.

Continuing this process as indicated in Fig. 2.27, we see that at time instant 4, the best paths to $s(4) = +1$ and $s(4) = -1$ have the same value of $s(1)$; indeed, they have the same values for $s(1)$, $s(2)$, and $s(3)$, namely $+1, +1$, and -1, respectively. Therefore, these values can all be definitely decided at time instant 4. The procedure continues until the end of the sequence is reached.

The optimum algorithm presented in this section is called the *maximum likelihood sequence estimator,* since it maximizes the conditional pdf in Eq. (2.7.2), called the *likelihood function,* which is the likelihood that the observed sequence would be received for all candidate input sequences. This algorithm not only outlines the optimum receiver for partial response signaling and channels with intersymbol interference, it is also important in the decoding of convolutional codes (see Sect. 3.4) and for the reception of sequences transmitted with continuous-phase modulation and coded modulation (see Chap. 4).

It has often been stated that there is a loss in signal-to-noise ratio (SNR) when partial response signaling is used in comparison with independent symbol transmission at the maximum rate (Lucky et al. 1968). However, this loss in SNR occurs only when optimum symbol-by-symbol detection is used at the receiver, and this loss does not occur when the receiver employs maximum likelihood sequence estimation to decode the received symbols (see Proakis (1989) for details concerning these performance comparisons).

2.8   Summary

The effects of channel noise in digital communications, including data transmission using analog modulation methods such as ASK, FSK, and PSK, have been developed based on the geometric approach introduced by Kotel'nikov (1947) and advanced by Wozencraft and Jacobs (1965). This approach unifies the analyses of digital communication systems and thus avoids the impression that communication theory is simply a collection of analytical techniques with little or no structure. Using this geometric approach, we have also been able to derive expressions for receivers that minimize the bit error probability and even gain insight into how one might choose the transmitted signals in order to optimize system performance. Particularly important signal constellations have been presented and comparisons of bit error probabilities given. The material presented in this chapter has had a profound effect upon digital communication system design for the last 20 years and is fundamental to understanding systems being developed today.

Problems

2.1   Use Eq. (2.2.13) to calculate all d_{ij} of interest for the signal set given by Eq. (2.2.16). Verify these distances by comparison with Fig. 2.1.

2.2   Write out the time-domain waveforms for the 16-level QAM signal set in Example 2.2.1. Specify all signal energies and the distances between each signal and its neighbors. Check by comparison with Fig. 2.2.

2.3   Verify that the signal set given by Eq. (2.2.21) can be represented by the signal space diagram in Fig. 2.4.

2.4   Specify an eight-phase PSK signal set in which one of the signals is $s_1(t) = A_c\sqrt{2/T}\cos[\omega_c t + 22.5°]$. Draw the corresponding signal space diagram.

2.5   Using $\varphi_1(t)$ in Eq. (2.2.22), specify an eight-level signal set over $0 \leq t \leq T/2$. Draw its signal space diagram.

2.6 Use Eqs. (2.2.22) and (2.2.23) to define a signal set corresponding to the signal space diagram in Fig. 2.2.

2.7 Given the waveforms $\{s_i(t), i = 1, 2, 3, 4\}$ shown in the figure below, use the Gram–Schmidt procedure to find appropriate signal vector representations.

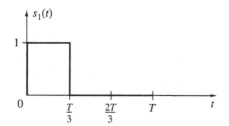

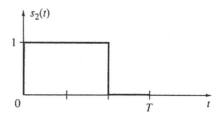

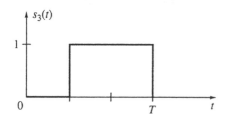

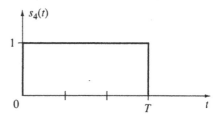

2.8 Renumber the signals in Fig. 2.5 as $s_1'(t) = s_4(t)$, $s_2'(t) = s_3(t)$, $s_3'(t) = s_2(t)$, and $s_4'(t) = s_1(t)$. Use the Gram–Schmidt procedure on this new signal set $\{s_i'(t), i = 1, 2, 3, 4\}$. Compare the results to Example 2.2.4.

2.9 Redefine the signals in the above figure as $s_1'(t) = s_3(t)$, $s_2'(t) = s_2(t)$, $s_3'(t) = s_4(t)$ and $s_4'(t) = s_1(t)$. Use the Gram–Schmidt procedure on this new "primed" signal set $\{s_i'(t), i = 1, 2, 3, 4\}$.

2.10 Consider the M-dimensional signal set given by the vectors

$$s_i = \left[s_{i1} \; s_{i2} \; \cdots \; s_{iM} \right], \quad i = 1, 2, \ldots, N,$$

where

$$s_{ij} = \begin{cases} +d \\ \text{or} \quad \text{for all } i, j. \\ -d \end{cases}$$

Sketch this signal set in two ($M = 2$) and three ($M = 3$) dimensions. How many signal vectors are there in M dimensions? Can you specify a set of time-domain waveforms that correspond to these signal vectors?

2.11 Specify the M-dimensional vectors for an orthogonal signal set. Sketch the signal sets in two and three dimensions. How many vectors are there in M dimensions?

2.12 Write signal vectors for the M-dimensional generalization of binary antipodal signals (called *biorthogonal* signals). Sketch these signal vectors in two and three dimensions. How many signals are there in M dimensions?

2.13 If each of the signal vectors in Problems 2.10, 2.11, and 2.12 has equal energy, say E_s, compare the distance between nearest neighbors for the three signal sets when $M = 3$.

2.14 Each member of the set of $N = M$ orthogonal signal vectors, denoted $\{s_i, i = 1, 2, \ldots, M\}$, is translated by $\mu = (1/M) \sum_{i=1}^{M} s_i$ to produce what is called the *simplex* signal set given by $s_i' = s_i - \mu, i = 1, 2, \ldots, M$. Sketch the simplex signal set in two and three dimensions. If $s_i \cdot s_j = E_s$ for all i, j for the orthogonal signal set, find $s_i' \cdot s_j'$ for all i, j for the simplex signal set.

2.15 In this problem and Problems 2.16 through 2.19, we prove that the vector r in Eq. (2.3.8) has all the "information" in $r(t)$ necessary to design an optimum receiver and that the pdf of the noise vector is given by Eq. (2.3.10). We begin by noting that the time-domain expression corresponding to Eq. (2.3.8) is

$$r'(t) \triangleq \sum_{j=1}^{M} r_j \varphi_j(t) = s_i(t) + n_o(t)$$

for some i, where $n_o(t) \triangleq \sum_{j=1}^{M} n_j \varphi_j(t)$ corresponds to $\mathbf{n}$. Now consider the difference signal $e(t) \triangleq r(t) - r'(t)$. Show that since $n(t)$ is a zero-mean Gaussian process, then $e(t)$ and $n_o(t)$ are zero-mean, jointly Gaussian processes.

2.16 Using the results of Problem 2.15, show that $e(t)$ and $n_o(t)$ are statistically independent.

2.17 Based on Eq. (2.4.3), show that if a receiver has $\mathbf{r}$ and $\mathbf{e}$ [a vector of time samples corresponding to $e(t)$] available, the conditional pdf $f_{e|r,s_i}$. is needed in the optimum receiver (see Problem 2.15).

2.18 Use Bayes' rule and Problem 2.16 to show that $\mathbf{e}$ is statistically independent of $\mathbf{r}$ and s_i, and hence only $\mathbf{r}$ is needed in the optimum receiver.

2.19 Use the results of Problems 2.15 and 2.16 to prove that $\mathbf{n}$ has the pdf in Eq. (2.3.10).

2.20 We consider the transmission of four messages over an AWGN channel with spectral density $\mathcal{N}_0/2$ W/Hz using the signal set specified by Eqs. (2.2.14) and (2.2.16). If $P[m_1] = \frac{1}{2}$, $P[m_2] = \frac{1}{4}$, and $P[m_3] = P[m_4] = \frac{1}{8}$, sketch the optimum decision regions.

2.21 The signal constellation in the figure below is used to send eight equally likely messages over a zero-mean AWGN channel with power spectral density $\mathcal{N}_0/2$ W/Hz. Determine and sketch the optimum decision regions.

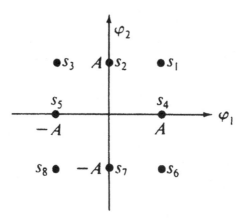

2.22 Repeat Problem 2.21 for the signal set in the figure below.

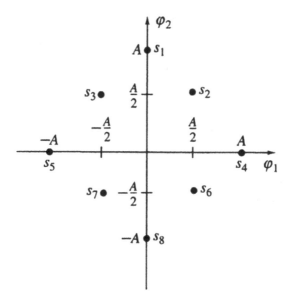

2.23 Derive the pdf in Eq. (2.4.34).

2.24 Use the approximation for $A_c/(\mathcal{N}_0/2) \gg 1$,

$$I_0\left(\frac{2A_c z}{\mathcal{N}_0}\right) \cong \frac{e^{2A_c z/\mathcal{N}_0}}{\sqrt{2\pi(2A_c z/\mathcal{N}_0)}},$$

to show that the pdf in Eq. (2.4.35) can be approximated as

$$f_{Z|s_2}(z|s_2) \cong \sqrt{\frac{z}{\pi A_c \mathcal{N}_0}}\, e^{-(z-A_c)^2/\mathcal{N}_0}.$$

Plot this pdf as a function of z, and hence validate the approximation in Eq. (2.5.19).

2.25 If the signal set in Problem 2.10, with equally likely signals, is used over an AWGN channel with power spectral density $\mathcal{N}_0/2$ W/Hz, what is $P[\mathcal{E}]$ for an optimum receiver?

2.26 Write an expression for $P[\mathcal{E}]$ if the M orthogonal signals in Problem 2.11 are used over an AWGN channel with power spectral density $\mathcal{N}_0/2$ W/Hz. Assume equally likely signals, each with energy E_s, and an optimum receiver.

2.27 Repeat Problem 2.26 for the simplex signal set.

2.28 Find an expression for the probability of error when the biorthogonal signals in Problem 2.12 are used to transmit equally likely messages over an AWGN channel with spectral density $\mathcal{N}_0/2$ W/Hz. Assume that each signal has energy E_s and an optimum receiver.

2.29 Plot $P[\mathcal{E}]$ for coherent and noncoherent ASK subject to a peak power constraint.

2.30 A received waveform $r(t)$ given by Eq. (2.3.1) is applied to the input of a linear filter with impulse response $h(t)$. Show that the filter which maximizes the signal-to-noise ratio (SNR) at its output is a matched filter. Let SNR $\triangleq s_i^2(t)/E\left[n^2(t)\right]$.

Hint: Use the Schwarz inequality,

$$\left|\int_{-\infty}^{\infty} A(\omega)B(\omega)d\omega\right|^2 \leq \int_{-\infty}^{\infty} |A(\omega)|^2 d\omega \int_{-\infty}^{\infty} |B(\omega)|^2 d\omega.$$

2.31 Draw block diagrams of the optimal correlation and matched filter receiver structure involving the set of possible transmitted waveforms $\{s_i(t), i = 1, 2, \ldots, N\}$. Compare to Fig. 2.13.

2.32 Repeat Example 2.7.1 for the case where the inputs to the duobinary system (after precoding) take on the four levels, $\pm 1, \pm\frac{1}{3}$.

2.33 Repeat Example 2.7.1 for binary-driven class 4 partial response.

Channel Coding

3

3.1   Introduction

Our focus in this chapter is on designing codes for the reliable transmission of digital
information over a noisy channel. The pertinent block diagram for this particular problem
is shown in Fig. 3.1. The binary message sequence presented to the input of the channel
encoder may be the output of a source encoder or the output of a source directly. For the
development in this chapter, we are only interested in the fact that the channel encoder
input is a binary sequence at a rate of R_s bits/s. The channel encoder introduces redundancy into the data stream by adding bits to the (input) message bits in such a way as
to facilitate the detection and/or correction of bit errors in the original binary message
sequence at the receiver. Perhaps the most familiar example of channel coding (for error
detection) to the reader is the addition of a parity check bit to a block of message bits.
For each k bits into the channel encoder, $n > k$ bits are produced at the channel encoder
output. Thus the channel coding process adds $n - k$ bits to each k-bit input sequence. As
a result, a dimensionless parameter called the *code rate,* denoted by R_c and defined by R_c
$= k/n =$ (number of bits in/number of bits out), is associated with the code. Since $k < n$,
and hence $R_c < 1$, the data rate at the channel encoder output is $R_T = R_s/R_c > R_s$ bps.
Thus the channel coding operation has increased the transmitted bit rate.

For the purposes of this chapter we assume that the modulator maps the coded bits in
a one-to-one fashion onto an analog waveform suitable for propagation over the channel,
and that the modulation process is totally separate from the coding operation. The modulator output is then transmitted over the channel and demodulated to produce a binary
sequence that serves as the input to the channel decoder. The channel decoder then generates a decoded binary sequence which hopefully is an accurate replica of the input binary
message sequence at the transmitter. The fact that the demodulator output is a binary
sequence means that the demodulator is making "hard decisions," and the subsequent

© The Author(s), under exclusive license to Springer Nature Switzerland AG 2023   97
J. D. Gibson, *Digital Communications*, Synthesis Lectures on Communications,
https://doi.org/10.1007/978-3-031-19588-4_3

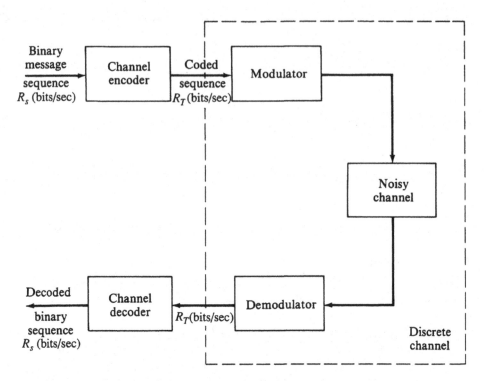

Fig. 3.1 Basic block diagram for the channel coding problem

decoding process is called *hard decision decoding*. The alternative to hard decisions is for the demodulator to pass on three or more level versions of its output to the decoder, which improves decoding accuracy over hard decisions (two-level information). The latter procedure is called *soft decision decoding* and generally increases receiver complexity. Soft decision decoding can substantially improve communication system performance; however, for our introductory treatment, we focus on hard decisions.

Since the input to the modulator is an R_T-bits/s binary sequence and the demodulator output is an R_T-bits/s binary sequence, it is common to lump the modulator and demodulator with the noisy channel and model the entire three block series as a discrete channel with binary inputs and binary outputs (shown by dashed lines in Fig. 3.1). Working with the lumped discrete (binary) channel, we now only need to consider channel encoders and decoders that have binary inputs and outputs. If this binary channel is memoryless and the probability of error for a 0 and a 1 are the same, we have the familiar binary symmetric channel (BSC).

The earliest types of codes often used in communication systems were block codes and convolutional codes. While in recent years, the applications of these codes have been more limited due to the emergence of turbo codes and low-density parity check codes,

the simple block and convolutional codes discussed here are predecessors and components of today's current coding schemes. Block and convolutional codes are discussed in
Sects. 3.2 and 3.4, respectively, with an emphasis on a few special cases and on forward
error correction (as opposed to error detection). In Sect. 3.3 we discuss a special class of
linear block codes, called *cyclic codes,* that are particularly easy to encode and decode
and hence have found a number of practical applications. In Sect. 3.5 we describe an error
control strategy called *automatic repeat-request* (ARQ), which relies on error detection
and retransmission rather than forward error correction. Some qualitative and quantita-
tive comparisons of code performance and implementation complexity are presented in
Sect. 3.6.

3.2 Block Codes

For the purposes of our current discussion, we refer to the input to the channel encoder
as V and the channel encoder output is denoted by W. We will call $V = \{v_1, v_2, \ldots\}$ the
message sequence. For block codes, the message sequence is partitioned into blocks of
fixed length k, indicated here by $\mathbf{v} = (v_1, v_2, \ldots, v_k)$. The channel encoder maps each
input message block $\mathbf{v}$ into an n-component output block called a *codeword* and denoted
by $\mathbf{w} = (w_1, w_2, \ldots, w_n)$, where $n > k$. Since we are only considering binary sequences,
there are 2^k possible messages and 2^k possible codewords, and the code is called an (n,
k) block code. We limit our attention to *linear* block codes, since linearity gives us the
structure needed to reduce encoding and decoding complexity.

A linear *systematic* block code has the additional structure that $n - k$ digits of a codeword are a linear combination of the components of the message block, while the other k
digits of the codeword are the k message digits themselves. Often, the two sets of digits
are grouped such that the first $n - k$ digits are a linear combination of the components of
the message block, while the last k digits are the k message digits themselves. Thus

$$w_1 = v_1 g_{11} + v_2 g_{21} + \cdots + v_k g_{k1}$$
$$w_2 = v_1 g_{12} + v_2 g_{22} + \cdots + v_k g_{k2}$$
$$\vdots$$
$$w_{n-k} = v_1 g_{1,n-k} + v_2 g_{2,n-k} + \cdots + v_k g_{k,n-k}$$
$$w_{n-k+1} = v_1$$
$$w_{n-k+2} = v_2$$
$$\vdots$$
$$w_n = v_k \tag{3.2.1}$$

where all additions in this chapter are considered to be modulo-2. Based on the set of equations in Eq. (3.2.1), we can construct a *generator matrix* given by

$$
\mathbf{G} = \begin{bmatrix} g_{11} & g_{12} & \cdots & g_{1,n-k} & 1 & 0 & \cdots \cdots & 0 \\ g_{21} & g_{22} & \cdots & g_{2,n-k} & 0 & 1 & 0 & \cdots & 0 \\ \vdots & & & & & \vdots \\ g_{k1} & g_{k2} & \cdots & g_{k,n-k} & 0 & \cdots \cdots & & 0 & 1 \end{bmatrix} \triangleq \begin{bmatrix} \mathbf{P} & \mathbf{I}_k \end{bmatrix}, \tag{3.2.2}
$$

where the $k \times n - k$ matrix $\mathbf{P}$ is defined as shown and $\mathbf{I}_k$ is a $k \times k$ identity matrix. (*Note:* Some authors may reverse $\mathbf{I}_k$ and $\mathbf{P}$ in $\mathbf{G}$. The performance is equivalent when used over a memoryless channel. In fact, any column permutation does not alter the code performance over a memoryless channel.). Thus the components of the codeword $\mathbf{w}$ in Eq. (3.2.1) can be produced by considering the k-tuple $\mathbf{v}$ as a row vector and performing the multiplication

$$
\mathbf{w} = \mathbf{v}\mathbf{G}. \tag{3.2.3}
$$

Thus, for a given generator matrix, we can obtain the codewords corresponding to all possible length-k sequences in a message block. We do not discuss how one tries to find a good $\mathbf{G}$ matrix here. It is not easy.

In addition to the $\mathbf{G}$ matrix, every linear block code has a related matrix called the *parity check matrix* $\mathbf{H}$, which for $\mathbf{G}$ in Eq. (3.2.2) is given by

$$
\mathbf{H} = \begin{bmatrix} \mathbf{I}_{n-k} & \mathbf{P}^T \end{bmatrix}, \tag{3.2.4}
$$

where $\mathbf{I}_{n-k}$ is the $n - k \times n - k$ identity matrix and $\mathbf{P}$ is as defined previously. The importance of $\mathbf{H}$ stems from the fact that

$$
\mathbf{w}\mathbf{H}^T = \mathbf{v}\begin{bmatrix} \mathbf{P} & \mathbf{I}_k \end{bmatrix} \begin{bmatrix} \mathbf{I}_{n-k} \\ \mathbf{P} \end{bmatrix} = \mathbf{v}\mathbf{P} + \mathbf{v}\mathbf{P} = \mathbf{0}, \tag{3.2.5}
$$

which simply says that any codeword multiplied by $\mathbf{H}^T$ is the $(n - k)$-length zero vector. This property is extremely useful for error detection and correction. Let $\mathbf{e}$ be an n-dimensional binary error vector added to $\mathbf{w}$ by the channel so that the channel output, and hence the channel decoder input is

$$
\mathbf{x} = \mathbf{w} + \mathbf{e}. \tag{3.2.6}
$$

Now the *syndrome* of the received vector $\mathbf{x}$ is defined as

$$
\mathbf{s} = \mathbf{x}\mathbf{H}^T. \tag{3.2.7}
$$

Clearly, if $\mathbf{x}$ is a codeword, then $\mathbf{s} = \mathbf{0}$, but if $\mathbf{x}$ is not a codeword, $\mathbf{s} \neq \mathbf{0}$. Thus an error is detected if $\mathbf{s} \neq \mathbf{0}$. The syndrome $\mathbf{s}$ is not dependent on which codeword is transmitted, and in fact, $\mathbf{s}$ depends only on the error vector $\mathbf{e}$. This can be demonstrated straightforwardly by substituting Eq. (3.2.6) into Eq. (3.2.7) and then using Eq. (3.2.5) to find that

$$\mathbf{s} = \mathbf{x}\mathbf{H}^T = \mathbf{w}\mathbf{H}^T + \mathbf{e}\mathbf{H}^T = \mathbf{e}\mathbf{H}^T. \tag{3.2.8}$$

Equation (3.2.8) implies that there are some *undetectable error patterns* that occur whenever $\mathbf{e}$ is a codeword. Since there are $2^k - 1$ nonzero codewords, there are $2^k - 1$ undetectable error patterns.

It would seem that Eq. (3.2.8) would allow us to find the error pattern $\mathbf{e}$ and hence implement the decoding operation at the receiver. However, Eq. (3.2.8) gives $n - k$ equations in n unknowns, the components of $\mathbf{e}$. Thus Eq. (3.2.8) has 2^k solutions, which implies that there are 2^k error patterns for each syndrome. How then does the decoder choose a single-error vector out of the 2^k possible error patterns for a given syndrome? The answer is that we select that error pattern which minimizes the probability of error as in Chap. 2. For independent, identically distributed channel inputs and a binary symmetric channel, the error vector that minimizes the probability of error is the error pattern with the smallest number of 1's. Note from Eq. (3.2.6) that $\mathbf{x}$ is the received codeword and once $\mathbf{e}$ is known, we can calculate the transmitted codeword as

$$\mathbf{w} = \mathbf{x} + \mathbf{e}, \tag{3.2.9}$$

since subtraction and addition modulo-2 are identical. Thus for each syndrome the decoder must have available the error vector with the fewest number of 1's out of the corresponding 2^k error patterns. The operation of the channel decoder therefore proceeds as follows:

Step 1. Calculate the syndrome of the received vector, that is, $\mathbf{s} = \mathbf{x}\mathbf{H}^T$.
Step 2. Find the error vector $\mathbf{e}$ with the fewest 1's corresponding to $\mathbf{s}$.
Step 3. Compute the decoder output as $\mathbf{w}' = \mathbf{x} + \mathbf{e}$.

The following example illustrates the many concepts presented thus far.

Example 3.2.1 A systematic linear block code called a (7,4) Hamming code has the generator matrix

$$\mathbf{G} = \begin{bmatrix} 1 & 1 & 0 & 1 & 0 & 0 & 0 \\ 0 & 1 & 1 & 0 & 1 & 0 & 0 \\ 1 & 1 & 1 & 0 & 0 & 1 & 0 \\ 1 & 0 & 1 & 0 & 0 & 0 & 1 \end{bmatrix}, \tag{3.2.10}$$

so that the $n = 7$-bit codewords for all possible $k = 4$-bit message blocks are shown in Table 3.1 and are calculated via Eq. (3.2.3). By inspection of Eq. (3.2.10),

$$\mathbf{P} = \begin{bmatrix} 1 & 1 & 0 \\ 0 & 1 & 1 \\ 1 & 1 & 1 \\ 1 & 0 & 1 \end{bmatrix}, \tag{3.2.11}$$

so that from Eq. (3.2.4), the parity check matrix of this code is

$$\mathbf{H} = \begin{bmatrix} 1 & 0 & 0 & 1 & 0 & 1 & 1 \\ 0 & 1 & 0 & 1 & 1 & 1 & 0 \\ 0 & 0 & 1 & 0 & 1 & 1 & 1 \end{bmatrix}. \tag{3.2.12}$$

Thus the validity of Eq. (3.2.5) can be demonstrated for any codeword in Table 3.1 using $\mathbf{H}$ in Eq. (3.2.12).

If we assume that the input message $\mathbf{v} = (1\ 0\ 1\ 0)$ is to be transmitted, the transmitted codeword is $\mathbf{w} = (0\ 0\ 1\ 1\ 0\ 1\ 0)$. For no channel errors, $\mathbf{e} = \mathbf{0}$ and

Table 3.1 The (7,4) hamming code

Message bits (v)	Codewords (w)
0 0 0 0	0 0 0 0 0 0 0
0 0 0 1	1 0 1 0 0 0 1
0 0 1 0	1 1 1 0 0 1 0
0 0 1 1	0 1 0 0 0 1 1
0 1 0 0	0 1 1 0 1 0 0
0 1 0 1	1 1 0 0 1 0 1
0 1 1 0	1 0 0 0 1 1 0
0 1 1 1	0 0 1 0 1 1 1
1 0 0 0	1 1 0 1 0 0 0
1 0 0 1	0 1 1 1 0 0 1
1 0 1 0	0 0 1 1 0 1 0
1 0 1 1	1 0 0 1 0 1 1
1 1 0 0	1 0 1 1 1 0 0
1 1 0 1	0 0 0 1 1 0 1
1 1 1 0	0 1 0 1 1 1 0
1 1 1 1	1 1 1 1 1 1 1

$$\mathbf{s} = \mathbf{w}\mathbf{H}^T = \begin{bmatrix} 0\ 0\ 1\ 1\ 0\ 1\ 0 \end{bmatrix} \begin{bmatrix} 1\ 0\ 0 \\ 0\ 1\ 0 \\ 0\ 0\ 1 \\ 1\ 1\ 0 \\ 0\ 1\ 1 \\ 1\ 1\ 1 \\ 1\ 0\ 1 \end{bmatrix} = \left(0\ 0\ 0 \right) = \mathbf{0}. \qquad (3.2.13)$$

If the channel produces the error pattern $\mathbf{e} = (1\ 0\ 0\ 0\ 0\ 0\ 0)$, then

$$\mathbf{x} = \mathbf{w} + \mathbf{e} = \left(1\ 0\ 1\ 1\ 0\ 1\ 0 \right) \qquad (3.2.14)$$

and

$$\mathbf{s} = \mathbf{x}\mathbf{H}^T = \begin{bmatrix} 1\ 0\ 1\ 1\ 0\ 1\ 0 \end{bmatrix} \begin{bmatrix} 1\ 0\ 0 \\ 0\ 1\ 0 \\ 0\ 0\ 1 \\ 1\ 1\ 0 \\ 0\ 1\ 1 \\ 1\ 1\ 1 \\ 1\ 0\ 1 \end{bmatrix} = \left(1\ 0\ 0 \right). \qquad (3.2.15)$$

To finish the decoding procedure, we need to find the error vector out of the $2^k = 2^4 = 16$ error vectors associated with the $\mathbf{s}$ that has the fewest 1's. We do this by letting $\mathbf{e} = (e_1\ e_2\ e_3\ e_4\ e_5\ e_6\ e_7)$ and writing out Eq. (3.2.8) as

$$1 = e_1 + e_4 + e_6 + e_7 \qquad (3.2.16a)$$

$$0 = e_2 + e_4 + e_5 + e_6 \qquad (3.2.16b)$$

$$0 = e_3 + e_5 + e_6 + e_7 \qquad (3.2.16c)$$

The 16 error patterns that satisfy these equations are shown in Table 3.2. Note that only one vector has a single 1, namely, $\mathbf{e} = (1\ 0\ 0\ 0\ 0\ 0\ 0)$, marked by an arrow. Thus for a BSC this is the most likely error pattern, and we compute the decoder output as [using $\mathbf{x}$ in Eq. (3.2.13)]

$$\mathbf{w}' = \mathbf{x} + \mathbf{e} = \left(0\ 0\ 1\ 1\ 0\ 1\ 0 \right),$$

which is the actual transmitted codeword. The corresponding message block from Table 3.1 is $\mathbf{v} = (1\ 0\ 1\ 0)$, and the decoder has correctly recovered the transmitted message block.

Table 3.2 Error patterns that satisfy Eqs. (3.2.16a, 3.2.16b, 3.2.16c)

e_1	e_2	e_3	e_4	e_5	e_6	e_7
1	1	0	1	1	1	0
1	0	0	0	0	0	0
1	0	1	0	1	1	1
1	1	0	1	1	1	0
0	1	1	1	1	1	1
0	0	0	0	1	1	0
0	0	1	0	0	0	1
1	1	1	1	0	0	1
0	1	0	0	1	0	1
0	1	0	1	0	0	0
0	0	1	1	1	0	0
0	0	0	1	0	1	1
0	0	1	0	0	0	1
1	1	1	0	1	0	0
1	0	1	1	0	1	0
1	1	0	0	0	1	1

When choosing a channel code, important questions that must be answered are: (1) How many errors does the code correct? and (2) How many errors can the code detect? The answers revolve around a parameter of the code called the minimum distance. To begin we define the *Hamming weight* or just weight of a codeword **w** as the total number of 1's in the codeword, and we denote the weight by $w(\mathbf{w})$. Therefore, the weight of the codeword $\mathbf{w} = (1\ 1\ 0\ 0\ 1\ 0\ 0)$ is $w(\mathbf{w}) = 3$. Using this idea, we next define the Hamming distance or just distance between any two codewords **w** and **v** as $w(\mathbf{w} + \mathbf{v})$, which is denoted by $d(\mathbf{w}, \mathbf{v})$. Thus, for **w** as just given and $\mathbf{v} = (0\ 0\ 1\ 0\ 0\ 1\ 1)$, $d(\mathbf{w}, \mathbf{v}) = w(\mathbf{w} + \mathbf{v}) = w[(1\ 1\ 1\ 0\ 1\ 1\ 1)] = 6$. The *minimum distance* for a code is finally given by

$$d_{\min} = \min\{d(\mathbf{w}, \mathbf{v}) = w(\mathbf{w} + \mathbf{v}),\ \text{over all codewords } \mathbf{w}, \mathbf{v} \text{ with } \mathbf{w} \neq \mathbf{v}\}. \qquad (3.2.17)$$

Now for a linear block code, the sum of any two codewords is also a codeword, so Eq. (3.2.17) can be rewritten as

$$d_{\min} = \min\{w(\mathbf{u}) \text{ for all codewords } \mathbf{u} \neq \mathbf{0}\} \triangleq w_{\min}. \qquad (3.2.18)$$

The quantity w_{min} is called the *minimum weight* of the code. From Eq. (3.2.18) we see that d_{min} for a code is simply the minimum weight of all nonzero codewords. Scanning the codewords in Table 3.1, it is evident that $d_{min} = 3$ for the (7,4) Hamming code.

Our minimum probability of error decoding scheme relies very heavily on the concept of how close a received vector is to a transmitted codeword. In fact, our decoder decodes the received vector as that codeword which is closest to the received vector in terms of Hamming distance. For this decoding to be unique, we need for the distance between any pair of codewords to be such that no received vector is the same distance from two or more codewords. This is achieved if for a *t*-error *correcting* code we have

$$d_{min} \geq 2t + 1. \tag{3.2.19}$$

The number of detectable errors for a given code also depends explicitly on d_{min}. Since d_{min} is the minimum distance between any pair of codewords, if the channel error pattern causes $d_{min} - 1$ or fewer errors, the received vector will not be a codeword, and hence the error is detectable. Thus a *t*-error *detecting* code only requires that

$$d_{min} \geq t + 1 \tag{3.2.20}$$

and comparing to Eq. (3.2.19), d_{min} can be smaller for *t*-error detecting codes than for *t*-error correcting codes. Since there are $2^k - 1$ nonzero codewords, there are $2^k - 1$ undetectable error patterns. (Why?) Further, since there are $2^n - 1$ nonzero error patterns, there are $2^n - 1 - (2^k - 1) = 2^n - 2^k$ detectable error patterns for an (n,k) linear block code. This is in contrast to the fact that the same linear block code has only $2^{n-k} - 1$ correctable error patterns.

Example 3.2.2 For the (7,4) Hamming code in Example 3.2.1, we know that $d_{min} = 3$. Thus this code can detect error patterns of $d_{min} - 1 = 2$ or fewer errors and can correct ($d_{min} = 3 = 2t + 1 \Rightarrow t = 1$) single errors. The total number of detectable error patterns is $2^n - 2^k = 2^7 - 2^4 = 112$. However, it can correct only seven single-error patterns.

Thus far we have avoided calculating the probability of error for a linear block code, and in general, this is not a trivial task. We can write an upper bound on the probability of a decoding error (block error) for a *t*-error correcting block code used over a BSC with independent probability of a bit error p by

$$P_B(\varepsilon) \leq \sum_{j=t+1}^{n} \binom{n}{j} p^j (1 - p)^{n-j}, \tag{3.2.21}$$

where the right side simply sums all possible combinations of $t + 1$ or more errors. Further consideration of error probabilities for linear block codes are not pursued here because of their complexity.

3.3   Cyclic Codes

A key issue in any application of error control codes is implementation complexity. A subset of linear codes, called *cyclic codes,* has the important property that encoders and decoders can be implemented easily using linearly connected shift registers. Given the code vector $\mathbf{w} = \left(w_1 \ w_2 \cdots w_n \right)$, we define a cyclic shift to the right of this code vector by one place as (Lin and Costello 1983)

$$\mathbf{w}^{(1)} = \left(w_n \ w_1 \quad w_2 \ldots w_{n-1} \right),$$

and a cyclic shift of i places by

$$\mathbf{w}^{(i)} = \left(w_{n-i+1} \ w_{n-i+2} \ldots w_n \ w_1 \ldots w_{n-i} \right).$$

A linear (n,k) code is called a cyclic code if every cyclic shift of a codeword is another codeword. Thus, for a cyclic code, if 1 0 0 1 0 1 1 is a codeword, so is 1 1 0 0 1 0 1.

The generator matrix of a cyclic code can take on a special form as illustrated by the generator matrix of a (7,4) cyclic code, which is

$$G' = \begin{bmatrix} 1\ 1\ 0\ 1\ 0\ 0\ 0 \\ 0\ 1\ 1\ 0\ 1\ 0\ 0 \\ 0\ 0\ 1\ 1\ 0\ 1\ 0 \\ 0\ 0\ 0\ 1\ 1\ 0\ 1 \end{bmatrix}. \tag{3.3.1}$$

Note that each row is a cyclic shift of an adjacent row. Exactly how this matrix is formed for a particular cyclic code is described in more detail later. Although the generator matrix is not in systematic form, it can be put in systematic form by elementary row operations. For example, keeping rows 1 and 2 intact, we replace row 3 by the sum of rows 1 and 3, and then replace row 4 by the sum of rows 1, 2, and 4, which yields the systematic generator matrix in Eq. (3.2.10). All sums are taken modulo-2.

We return to the specification of generator matrices and their corresponding parity check matrices later. To facilitate the development of encoder and decoder shift register implementations for cyclic codes, we present an alternative development of cyclic codes in terms of a polynomial representation. The development here is adapted from Lin and Costello (1983) and Bertsekas and Gallager (1987). Let

$$w(D) = w_n D^{n-1} + w_{n-1} D^{n-2} + \cdots + w_2 D + w_1 \tag{3.3.2}$$

be the polynomial in D corresponding to the transmitted codeword, and let

$$v(D) = v_k D^{k-1} + v_{k-1} D^{k-2} + \cdots + v_2 D + v_1 \tag{3.3.3}$$

be the polynomial corresponding to the message bits. The codeword polynomial is expressible as the product of $v(D)$ and the generator polynomial $g(D)$, so

$$w(D) = v(D)g(D). \tag{3.3.4}$$

The generator polynomial of an (n,k) cyclic code is a factor of the polynomial $D^n + 1$; in fact, any polynomial of degree $n - k$ that is a factor of $D^n + 1$ generates an (n,k) cyclic code. For example, for the (7,4) linear code, we have

$$D^7 + 1 = (1 + D)(1 + D^2 + D^3)(1 + D + D^3). \tag{3.3.5}$$

There are two polynomials of degree $n - k = 3$ that generate cyclic codes. The factor $1 + D + D^3$ is known to generate a single error-correcting cyclic code of minimum distance 3. The generator matrix for this generator polynomial is the one in Eq. (3.3.1), where the coefficients of increasing powers of D appear in the first row of the matrix starting with the D^0. Subsequent rows are cyclic shifts of the first row.

To get the systematic form of the generator matrix starting with the generator polynomial, we divide $D^{n-k+i-1}$ by $g(D)$ for $i = 1, 2, \ldots, k$, which yields

$$D^{n-k+i-1} = a_i(D)g(D) + \mathbf{g}_i(D) \tag{3.3.6}$$

or

$$D^{n-k+i-1} + \mathbf{g}_i(D) = a_i(D)g(D), \tag{3.3.7}$$

where we write

$$\mathbf{g}_i(D) = g_{i1} + g_{i2}D + \cdots + g_{i,n-k}D^{n-k-1}. \tag{3.3.8}$$

The coefficients $\{g_{ij}; i = 1, 2, \ldots, k, j = 1, \ldots, n - k\}$ are the coefficients in the systematic generator matrix of Eq. (3.2.2).

To encode in systematic form using the polynomial notation, we observe that the code vector polynomial must equal the message vector polynomial multiplied by D^{n-k} and summed with a parity check polynomial of order $n - k - 1$. Thus

$$w(D) = D^{n-k}v(D) + p(D), \tag{3.3.9}$$

where

$$p(D) = p_1 + p_2D + \cdots + p_{n-k}D^{n-k-1}. \tag{3.3.10}$$

Since $w(D) = a(D)g(D)$, $p(D)$ is just the remainder when we divide $D^{n-k}v(D)$ by $g(D)$. Expanding Eq. (3.3.9), we obtain

$$w(D) = D^{n-k}v(D) + p(D)$$
$$= v_k D^{n-1} + v_{k-1}D^{n-2} + \cdots + v_1 D^{n-k}$$
$$+ p_{n-k}D^{n-k-1} + \cdots + p_2 D + p_1 \qquad (3.3.11)$$

so that

$$\mathbf{w} = \left(p_1\ p_2\ \cdots\ p_{n-k}\ v_1\ v_2\ \cdots\ v_k \right). \qquad (3.3.12)$$

The first $n-k$ symbols are parity check bits and the last k digits are the information bits.

Example 3.3.1 Starting with the generator polynomial $g(D) = 1+D+D^3$ of the (7,4) linear code, we wish to encode the message vector $\mathbf{v} = \left(1\ 0\ 1\ 0 \right)$ using the systematic version of the code. Thus we need to find $p(D)$ as the remainder when dividing $D^3 \left[D^2 + 1 \right] = D^5 + D^3$ by $g(D)$. Using synthetic division, we get $p(D) = D^2$ and $w(D) = D^5 + D^3 + D^2$ and the transmitted codeword from Eq. (3.3.12) is $\mathbf{w} = \left(0\ 0\ 1\ 1\ 0\ 1\ 0 \right)$, which agrees with the result in Sect. 3.2.

We can also start with $g(D)$ and form the systematic generator matrix based on the approach in Eqs. (3.3.6)–(3.3.8). $D^{n-k+i-1} = D^3, D^4, D^5, D^6$, for $i = 1, 2, 3$, and 4, respectively, so

$$\mathbf{g}_1(D) = 1 + D$$
$$\mathbf{g}_2(D) = D + D^2$$
$$\mathbf{g}_3(D) = 1 + D + D^2$$
$$\mathbf{g}_4(D) = 1 + D^2$$

and the generator matrix in Eq. (3.2.10) follows from Eq. (3.2.2).

The polynomial operations for decoding a cyclic code are rather straightforward. Let the polynomial corresponding to the received vector be

$$x(D) = w(D) + e(D), \qquad (3.3.13)$$

where $e(D)$ is the error vector polynomial. Dividing $x(D)$ by $g(D)$ gives

$$x(D) = b(D)g(D) + s(D) \qquad (3.3.14)$$

with $s(D)$ the syndrome polynomial. Since $w(D) = a(D)g(D)$, we can combine Eqs. (3.3.13) and (3.3.14) to produce a relationship between the syndrome polynomial and the error pattern polynomial,

$$e(D) = (a(D) + b(D))g(D) + s(D). \qquad (3.3.15)$$

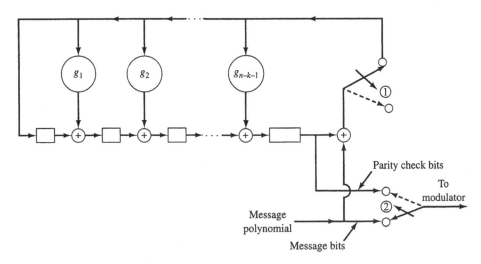

Fig. 3.2 Encoding of a cyclic code based on the generator polynomial. From J. G. Proakis, *Digital Communications,* 2nd ed., New York: McGraw-Hill, Inc., 1989. Reproduced with permission of McGraw-Hill, Inc

If there are no errors, $e(D) = 0$. Since we know that $g(D)$ divides $w(D)$, we assume that $b(D) = a(D)$. Thus, from Eq. (3.3.15), this yields $s(D) = 0$.

Referring to Eqs. (3.3.9)–(3.3.11) for encoding and Eqs. (3.3.13) and (3.3.14) for decoding, we see that both of these operations require the division of one polynomial by another polynomial. Fortunately, this operation is implementable with a linear feedback shift register circuit. To be more specific, recall that encoding requires multiplication of $v(D)$ by D^{n-k}, division of $D^{n-k}v(D)$ by $g(D)$ to get the remainder, and the formation of $D^{n-k}v(D) + p(D)$. Figure 3.2 shows a general circuit for accomplishing this. With switch ① in the up position and switch ② in the down position, the message bits are shifted in. Immediately, after the last of these bits have been shifted in, the shift register contents are the coefficients of $p(D)$. Switch ① is moved to the lower position and switch ② to the up position, and the contents of the shift register are sent to the modulator. In this manner, **w** in Eq. (3.3.12) is formed.

Example 3.3.2 A linear feedback shift register connection for the (7,4) linear code is shown in Fig. 3.3, where the shift register contents are for $v(D) = D^2 + 1$. Table 3.3 shows the contents of the shift register as the message is shifted in. The final output of the circuit is the 7-bit codeword found in Example 3.3.1.

A linear shift register feedback connection for calculating the syndrome based on $g(D)$ is shown in Fig. 3.4. The initial contents of the shift register are all zeros. The components

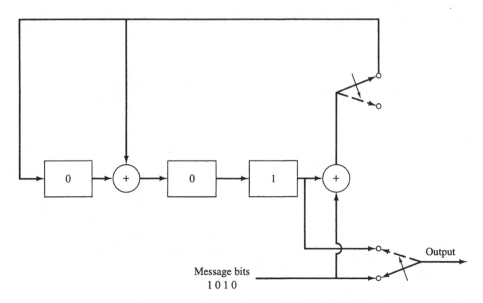

Fig. 3.3 Encoding circuit for a (7,4) cyclic code. From J. G. Proakis, *Digital Communications,* 2nd ed., New York: McGraw-Hill, Inc., 1989. Reproduced with permission of McGraw-Hill, Inc

Table 3.3 Shift register contents for the encoder of Fig. 3.3

Input	Shift register contents
	0 0 0
0	0 0 0
1	1 1 0
0	0 1 1
1	0 0 1 $= (p_1 p_2 p_3)$

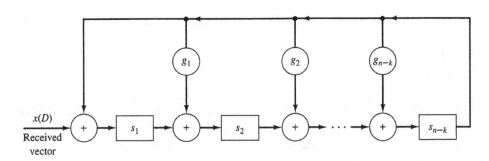

Fig. 3.4 Shift register connection for obtaining the syndrome

of $x(D)$ are shifted in one bit at a time, and after the last bit is shifted in, the contents of the shift register specify the syndrome.

Example 3.3.3   A syndrome calculation circuit for the (7,4) cyclic code based on $g(D) = D^3 + D + 1$ is shown in Fig. 3.5. The shift register contents as $x(D) = 1 + D^2 + D^3 + D^5$ is shifted in are listed in Table 3.4.

The simple fact that $g(D)$ divides $w(D)$ and the assumption that if $g(D)$ divides $x(D)$ the received codeword is accepted as being correct, allow a number of conclusions to be reached concerning the error-detecting capability of a cyclic code. If there is a single bit error in position i, then $e(D) = D^i$. For this error to not be detected, $g(D)$ must divide $e(D)$. However, if $g(D)$ has more than one term, it cannot divide D^i, and all single errors are detectable.

For a double bit error in the ith and jth positions, we have

$$e(D) = D^i + D^j = D^i\left(1 + D^{j-i}\right),$$

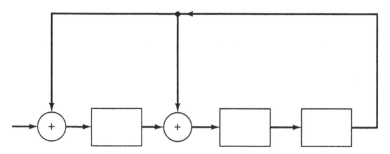

Fig. 3.5   Circuit for calculating the syndrome of the (7,4) code with $g(D) = D^3 + D + 1$

Table 3.4 Syndrome calculation corresponding to Fig. 3.5

Received	Shift register contents
	0 0 0
0	0 0 0
1	1 0 0
0	0 1 0
1	1 0 1
1	0 0 0
0	0 0 0
1	1 0 0 = $(s_1 s_2 s_3)$

Table 3.5 Generator polynomials for CRC code standards

Code	Generator polynomial
CRC-12	$1 + D + D^2 + D^3 + D^{11} + D^{12}$
CRC-16	$1 + D^2 + D^{15} + D^{16}$
CRC-CCITT	$1 + D^5 + D^{12} + D^{16}$
CRC-32	$1 + D + D^2 + D^4 + D^5 + D^7 + D^8 + D^{10} + D^{11} + D^{12} + D^{16} + D^{22} + D^{23}$ $+ D^{26} + D^{32}$

assuming that $i < j$. If $g(D)$ has three or more terms, neither component will be divisible by $g(D)$ and all double errors will be detected.

Some statements can also be made concerning burst error detection, that is, a sequence of consecutive bit errors. If the burst is of length $n - k$ or less, we can write $e(D)$ as

$$e(D) = D^i B(D), \qquad (3.3.16)$$

where $B(D)$ is of degree $n - k - 1$ or less and $0 \leq i \leq n - 1$. The generator polynomial is of degree $n - k$, and if it has more than one term, it cannot divide either factor of Eq. (3.3.16). Therefore, all bursts of length $n - k$ or less are detectable. Other burst error-detecting capabilities are left to the problems (Problems 3.21 and 3.22).

Some cyclic codes are called *cyclic redundancy check* (CRC) *codes* and a few have been selected as standards for networking applications. The generator polynomials for these codes are given in Table 3.5.

3.4 Convolutional Codes

Convolutional codes are important channel codes for many applications, and the encoding and decoding operations for convolutional codes differ quite substantially from those described for linear block codes in the preceding section. The nomenclature and structure of convolutional codes are most simply (and most commonly) developed by example. Thus we begin by considering the convolutional code in Fig. 3.6 with binary input sequence $\mathbf{v} = (v_1 \ v_2 \ v_3 \ \cdots)$ and binary output sequence $\mathbf{w} = (w_1 \ w_2 \ w_3 \ \cdots)$. The blocks labeled "$D$" represent unit delays and the $\oplus$ represent summations. The output sequence is obtained by alternately sampling the upper and lower summer outputs. The encoder inputs and outputs are not blocked as in Sect. 3.2 but are "semi-infinite" binary sequences. For the input sequence $\mathbf{v} = \left(1\,0\,1\,1\,0\,1\,0\,0 \cdots \right)$, where the three dots indicate a repetition of zeros, the output sequence can be found from Fig. 3.6 to be (assuming zero initial conditions)

$$\mathbf{w} = \left(1\,1\,0\,1\,0\,0\,1\,0\,1\,0\,0\,0\,0\,1\,1\,1\,0\,0 \cdots \right).$$

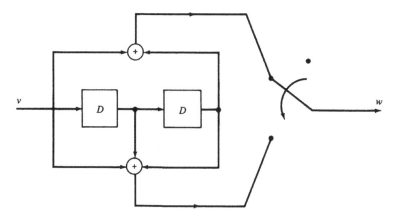

Fig. 3.6 An $(n,k) = (2, 1)$ binary convolutional encoder

For each input bit in $\mathbf{v}$ there are two output bits in $\mathbf{w}$, so the code is said to have rate $R = k/n = \frac{1}{2}$. Our development in this section is limited to rate $1/n$ convolutional codes (see Problem 3.35 for an example of a higher rate code).

There are several alternative ways of drawing a convolutional encoder that appear in the literature. One alternative is illustrated in Fig. 3.7, where delay elements are not indicated, but storage locations are shown. Some authors say that the diagram in Fig. 3.6 is a two-stage shift register (by counting delays), while others claim that Fig. 3.7 represents a three-stage shift register by counting storage locations. This is only nomenclature, however, and either diagram is easy to work with. Another point of confusion in convolutional coding terminology is the definition of *constraint length*. We define the constraint length of a convolutional code to be equal to the number of storage locations or digits available as inputs to the modulo-2 summers, and denote it by v. Several other definitions of constraint length are employed in the literature, all of which are related to our definition of v in some manner.

The encoder output is an interleaving of the outputs of the upper and lower summer outputs. If the encoder input is an impulse, $\mathbf{v} = \left(1\,0\,0\,\cdots \right)$, the impulse responses of the upper and lower branches are

$$\mathbf{w}^{(1)} = \left(1\,0\,1\,0\,0\,\cdots \right) \tag{3.4.1}$$

and

$$\mathbf{w}^{(2)} = \left(1\,1\,1\,0\,0\,\cdots \right). \tag{3.4.2}$$

These impulse responses are usually given the special name *generator sequences* and denoted by $\mathbf{g}^{(1)} = \mathbf{w}^{(1)}$ in Eq. (3.4.1) and $\mathbf{g}^{(2)} = \mathbf{w}^{(2)}$ in Eq. (3.4.2). For any given

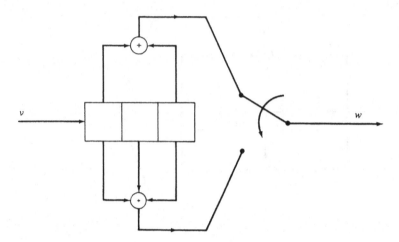

Fig. 3.7 Alternative diagram for an $(n,k) = (2, 1)$ binary convolutional encoder

encoder input $\mathbf{v}$, the two outputs $\mathbf{w}^{(i)}$, $i = 1, 2$ can be found by discrete-time convolution for all time instants $m \geq 1$ as

$$\mathbf{w}^{(i)} = \mathbf{v} * \mathbf{g}^{(i)} = \sum_{j=0}^{\nu-1} v_{m-j} g_{j+1}^{(i)} = v_m g_1^{(i)} + v_{m-1} g_2^{(i)} + \cdots + v_{m-\nu+1} g_\nu^{(i)} \qquad (3.4.3)$$

with $v_{m-j} \triangleq 0$ for $m - j \leq 0$. The encoder output sequence is then $w = \left(w_1^{(1)} w_1^{(2)} \ w_2^{(1)} w_2^{(2)} w_3^{(1)} w_3^{(2)} \ldots \right)$. A generator matrix for convolutional codes can be obtained by interleaving the generator sequences to form a row and then creating subsequent rows by shifting n columns to the right. All unspecified components are zero. The generator matrix for the code in Figs. 3.6 and 3.7 is

$$\mathbf{G} = \begin{bmatrix} g_1^{(1)} \ g_1^{(2)} \ g_2^{(1)} \ g_2^{(2)} \ g_3^{(1)} \ g_3^{(2)} \ \cdots \ g_\nu^{(1)} \ g_\nu^{(2)} \\ \cdots \ \cdots \ g_1^{(1)} \ g_1^{(2)} \ g_2^{(1)} \ g_2^{(2)} \ \cdots \cdots \ \cdots \ g_\nu^{(1)} \ g_\nu^{(2)} \\ \ddots \end{bmatrix}$$

$$= \begin{bmatrix} 1\ 1\ 0\ 1\ 1\ 1\ 0\ 0 \cdots \cdots \cdots \cdots \\ \quad 1\ 1\ 0\ 1\ 1\ 1\ \ 0\ \ 0 \ \cdots \cdots \cdots \\ \quad\quad 1\ 1\ 0\ 1\ \ 1\ \ \ 1\ \ \ 0\ \ \ 0 \ \cdots \end{bmatrix}. \qquad (3.4.4)$$

The encoder output can be written in terms of this generator matrix and the encoder input treated as a row vector as

$$\mathbf{w} = \mathbf{v}\mathbf{G} \qquad (3.4.5)$$

so for $\mathbf{v} = \Big(1\,0\,1\,1\,0\,1\,0\,0\,\cdots\Big)$ as before,

$$\mathbf{w} = \mathbf{v}G = \Big(1\,1\,0\,1\,0\,0\,1\,0\,1\,0\,0\,0\,0\,1\,1\,1\,0\,0\,\cdots\Big).$$

Of course, $\mathbf{G}$ is different from the linear block code case since it is a semi-infinite matrix here.

Graphical displays of convolutional codes have proven invaluable over the years for their understanding and analysis. A particularly useful graphical presentation is the *code tree*. A code tree is created by assuming zero initial conditions for the encoder and considering all possible encoder input sequences. For a 0 input digit a tree branch is drawn by moving upward, drawing a horizontal line and labeling the line with the encoder output. For a 1 input a tree branch is drawn by moving down and drawing a horizontal line labeled with the encoder output. The same process is repeated at the end of each horizontal line indefinitely. The code tree for the convolutional code in Figs. 3.6 and 3.7 is shown in Fig. 3.8. Note that the output $\mathbf{w}$ in Eq. (3.4.5) can be obtained using the code tree and the given encoder input by stepping up for an input 0 and down for an input 1 and reading off the two-digit branch labels. To find all of $\mathbf{w}$ in Eq. (3.4.5), the code tree needs to be extended. The tree diagram (code tree) thus allows the encoder output to be found very easily for a given input sequence. Unfortunately, for l input bits, the tree has 2^l branches, so that as l, the number of input bits, gets large, the code tree has an exponentially growing number of branches.

The code tree can be redrawn into a more manageable diagram called a *trellis* by noting that after v input bits, the tree becomes repetitive. Each of the nodes (or branching points indicated by black dots) in Fig. 3.8 is labeled by one of the numbers 0, 1, 2, or 3. These numbers indicate the two most recent encoder input bits by the correspondence $0 \leftrightarrow 00$, $1 \leftrightarrow 01$, $2 \leftrightarrow 10$, and $3 \leftrightarrow 11$, and are usually called *states*. All like-numbered nodes after depth $2(v-1)$ in the tree have identical output digits on the branches emanating from them. Therefore, these nodes can be merged. By merging like-numbered nodes after depth $2(v-1)$ in the tree, we obtain a trellis, as shown in Fig. 3.9. As in the tree, an upper branch out of a node is taken for a 0 input digit and the lower branch is taken for a 1 input bit. The branch labels are the output bits. Note that after all states are reached, the trellis is repetitive. For a given input sequence, we can find the output by tracing through the trellis, but without the exponential growth in branches as in the tree.

Just as the encoding operation for convolutional codes is quite different from the encoding operation for linear block codes, the decoding process for convolutional codes proceeds quite differently. Since we can represent a transmitted codeword for a convolutional code as a path through a trellis, the decoding operation consists of finding that path through the trellis which is "most like" the received binary sequence. As with the linear block codes, we are interested in hard decision decoding and a decoder that minimizes the probability of error, and therefore, for a given received binary vector, the decoder finds

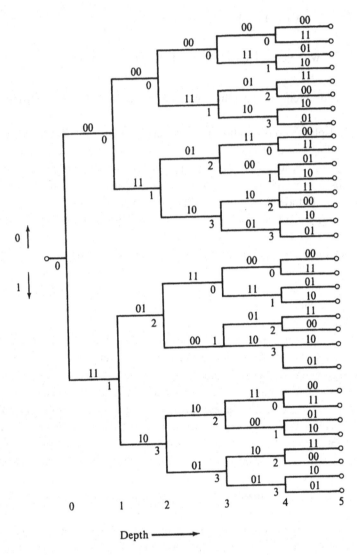

Fig. 3.8 Code tree for the binary convolutional encoder in Figs. 3.6 and 3.7

that path through the trellis which has minimum Hamming distance from the received sequence. Given a long received sequence of binary digits and a trellis similar to that in Fig. 3.9, it would seem quite a formidable task to search all possible paths for the best path. However, there exists an iterative procedure called the Viterbi algorithm that greatly simplifies matters. This algorithm is a special case of *forward dynamic programming* and relies on the "principle of optimality" (see Sect. 2.7). As applied to our particular problem, this principle states that the best (smallest Hamming distance) path through the trellis

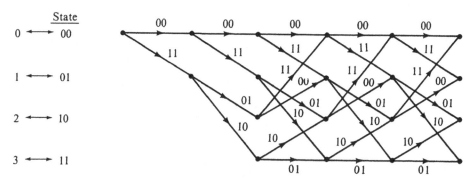

Fig. 3.9 Trellis corresponding to the tree in Fig. 3.8

that includes a particular node necessarily includes the best path from the beginning of the trellis to this node. What this means to us is that for each node in the trellis we need only retain the single best path to a node, thus limiting the number of retained paths at any time instant to the number of nodes in the trellis at that time. Therefore, for the trellis in Fig. 3.9, no more than four paths are retained at any time instant. The retained paths are called *survivors,* and the nodes correspond to *states* as shown in the figure. The basic steps in the Viterbi algorithm are as follows:

> Step 1. At time instant j, find the Hamming distance of each path entering a node by adding the Hamming distance required in going from the node(s) at time $j - 1$ to the node at time j with the accumulated Hamming distance of the survivor into the node at time $j - 1$.
> Step 2. For each node in the trellis, discard all paths into the node except for the one with the minimum total Hamming distance—this is the survivor.
> Step 3. Go to time instant $j + 1$ and repeat steps (1) and (2).

Use of the Viterbi algorithm can be clarified by a concrete example.

Example 3.4.1 We assume that the rate-$\frac{1}{2}$ binary convolutional code represented by the encoders in Figs. 3.6 and 3.7 and the trellis in Fig. 3.9 is used and that the transmitted codeword is the all-zero sequence $\mathbf{w} = \left(0\,0\,0\,0\,0\,0\,0 \cdots \right)$. The received codeword is assumed to be $\mathbf{x} = \left(1\,0\,1\,0\,0\,0 \cdots \right)$ and we wish to use the Viterbi algorithm to decode this received sequence.

The decoding operation is indicated by a series of incomplete trellises in Fig. 3.10. Through time instant 2, no paths need be discarded, since the trellis is just "fanning out." The four possible paths with the Hamming distance into each node listed above the node are shown in Fig. 3.10a. These distances are calculated by comparing the received bits with the path labels. Thus the Hamming distance associated with state 2 at time instant 2

Fig. 3.10 Viterbi decoding for
Example 3.4.1

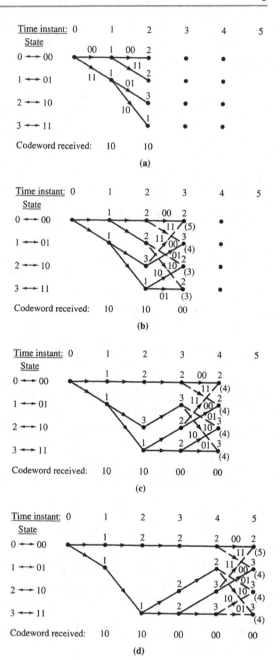

is $\mathbf{w}\left[\left(1\ 0\ 1\ 0\right)+\left(1\ 1\ 0\ 1\right)\right] = \mathbf{w}\left[\left(0\ 1\ 1\ 1\right)\right] = 3$, as shown. We now extend each of the paths to the next level and calculate Hamming distances as demonstrated in Fig. 3.10b. The paths to be discarded are shown dashed and their corresponding Hamming distances are given in parentheses below the node. The minimum Hamming distance to each node is written above the node, and the retained paths (survivors) are the solid lines. The survivors are extended and the calculations repeated in Fig. 3.10c, d. After time instant 5 in Fig. 3.10d, it is evident that the decoder output will be $\mathbf{y} = \left(0\ 0\ 0\ 0\ 0\ 0\ \cdots\right)$, since the total Hamming distance to state 0 at time instant 5 is 2, while the distance to all other nodes is 3. Since subsequent decisions cannot reduce the path distance, the all-zero sequence is the decoder output. In this instance, the decoder correctly decoded the received codeword. This is not always the case, since uncorrectable error patterns can occur for convolutional codes just as for block codes.

The Viterbi algorithm has the drawback that the number of decoding calculations is proportional to $2^{\nu-1}$, where ν is the constraint length. As ν gets large, the computational load of Viterbi decoding becomes prohibitive. Additionally, the Viterbi decoder requires the same decoding effort, whether the channel is very noisy or relatively error-free. For these reasons other decoders for convolutional codes have been developed. Examination of these other decoding algorithms is beyond the intended scope of this book. Other important topics not covered here are probability of error calculations for convolutional codes and soft decision decoding, which is very useful with convolutional codes. The reader should consult the references for additional information (Lin and Costello 1983).

3.5 Automatic Repeat-Request Systems

Sections 3.2 and 3.4 are concerned primarily with what is called *forward error correction* (FEC), that is, codes which try to detect *and* to correct channel errors. The possibility of using linear block codes for error detection alone is also mentioned briefly in Sects. 3.2 and 3.3, where it is noted that error detection is easier than error correction. This is because for each $(n - k)$-length syndrome there are 2^k error patterns that could have generated this syndrome, but only one of the error patterns can be corrected. Thus there are $2^{n-k} - 1$ correctable error patterns, but as demonstrated in Sect. 3.2, there are $2^n - 2^k$ detectable error patterns. From Example 3.2.2 for the (7,4) Hamming code, there are 112 detectable error patterns but only $2^{7-4} - 1 = 7$ correctable error patterns. If error detection is simpler than error correction, how do we make use of the fact that we have detected an error? *Automatic repeat-request* (ARQ) *systems* employ strategies that use error detection information to improve communication system performance.

Here we discuss three basic ARQ strategies: (1) the stop-and-wait ARQ, (2) the go-back-N ARQ, and (3) the selective-repeat ARQ. As for any communication system, these strategies differ in complexity and performance. Generally, ARQ systems employ a good error detecting code and compute the syndrome of the received vector. If the syndrome is zero, the received vector is taken to be correct, and the receiver notifies the transmitter over a reverse or return channel that the transmitted vector has been correctly received by sending an acknowledgment (ACK). The reverse channel is assumed to be low speed and accurate (no errors). When the syndrome is nonzero, an error is detected in the received vector, and the transmitter is informed of this occurrence by the receiver sending a negative acknowledgment (NAK) through the return channel. When the transmitter receives the NAK, it retransmits the erroneously received vector. The procedure is repeated until all received vectors have a zero syndrome (no error is detected). The specific ARQ strategies are variations on this theme.

The stop-and-wait ARQ method has the transmitter send a codeword to the receiver and then not send another codeword until a positive acknowledgment, ACK, is returned by the receiver. If NAK is returned to the transmitter, the codeword is retransmitted until ACK is received by the transmitter. After reception of an ACK, the next codeword is transmitted. This stop-and-wait ARQ scheme is obviously inefficient because of the time spent waiting for an acknowledgment. One suggestion for overcoming this idle time inefficiency is to use very long block length codes, but this has the disadvantage of increasing retransmissions, since a longer block is more likely to contain errors.

The go-back-N ARQ system continuously transmits codewords rather than waiting for an acknowledgment after each received vector. The quantity $N - 1$ is the number of codewords transmitted during the time it takes for a vector to reach the receiver and for the acknowledgment of the vector to be returned to the transmitter. This time is called the *round-trip delay*. When an NAK is returned, the transmitter retransmits the erroneously received codeword *and* the $N - 1$ immediately following codewords. Thus a buffer is not needed at the receiver since the $N - 1$ vectors following an erroneous received vector are discarded irrespective of whether they are correctly received. This go-back-N ARQ method efficiently utilizes the forward channel because of its continuous transmission, but it may not be effective at delivering usable data to the receiver if the transmitted bit rate is high and the round-trip delay relatively large. This is because N will be large in this case, and too many error-free codewords are retransmitted after an error is detected at the receiver.

The selective-repeat ARQ scheme improves upon the preceding ARQ systems by continuously transmitting codewords and retransmitting only erroneously received codewords. Therefore, as for the go-back-N system, codewords are continuously transmitted and those codewords for which an NAK is returned are retransmitted. However, the selective-repeat ARQ method *only* resends codewords for which an NAK is returned. For received vectors to be released to the user in the correct order, a buffer is now required at the receiver. Correctly received vectors are continuously released until a codeword is received in error. Subsequently received correct codewords are then stored at the receiver until the incorrect codeword is received without error. The entire string of codewords is then released from the receiver buffer to the user in the correct order. The receiver buffer must be sufficiently long to prevent buffer overflow for this system.

The primary indicator of ARQ performance is what is called the *throughput,* which is defined to be the ratio of the average number of message digits delivered to the user per unit time to the maximum number of digits that could be transmitted per unit time. For an (n,k) linear block code and with P being the probability of a received vector being accepted, the throughput of the selective-repeat system is (Lin and Costello 1983)

$$\eta_{sr} = \frac{k}{n}P \tag{3.5.1}$$

and the throughput of the go-back-N scheme is

$$\eta_{gbN} = \frac{P}{P + (1-P)N}\frac{k}{n} \tag{3.5.2}$$

To write the expression for the throughput of the stop-and-wait method, we define the time spent waiting for an acknowledgment as τ_d and the transmitted data rate in bits/s as r. The throughput of the stop-and-wait scheme can now be written as

$$\eta_{sw} = \frac{P}{1 + (r\tau_d/n)}\frac{k}{n}. \tag{3.5.3}$$

Clearly, η_{sr} is greater than both η_{gbN} and η_{sw}. Note, however, that the expression for η_{sr} in Eq. (3.5.1) assumes an infinite buffer at the receiver. More detailed analysis with a reasonable finite buffer size of say, N, reveals that the throughput of the selective-repeat system is still significantly better than the other two (Lin and Costello 1983).

3.6   Code Comparisons

The linear block codes in Sect. 3.2 and the convolutional codes in Sect. 3.4, along with the decoding algorithms presented in these sections, are but a very few of a host of channel codes and their decoding algorithms that have been investigated and described in the literature. For a variety of reasons, the performance evaluation of channel codes can be a difficult task that is certainly beyond the scope of this book. We collect here some performance data concerning both block and convolutional codes to give the reader an idea of what can be expected.

Of course, an often used indicator of relative performance for linear block codes is the minimum distance d_{min}, since the number of errors that can be detected and corrected is proportional to d_{min}. There is a related quantity that can be defined for convolutional codes called the *free distance,* d_{free}, which is the minimum distance between any two codewords in a code. Because of the exponential growth with constraint length v of the number of states and possible codewords in a convolutional code, the calculation of d_{free} for a code can be a difficult task. In any event, when d_{free} is known, the larger d_{free}, the better the code performance.

Another shortcut often taken to ascertaining how well a class of codes performs is to devise coding bounds. These bounds are primarily of two types, bounds on $d_{\min}$ and bounds on performance. Two bounds on $d_{\min}$, the Hamming bound and the Plotkin bound, give the maximum possible $d_{\min}$ for a specific code rate and code block length. Another bound, called the Gilbert bound, gives a lower bound on $d_{\min}$ for the best code. The other types of bounds, performance bounds, are generally of the type called *random coding bounds*. Bounds in this class bound the *average* performance of all codes with a certain structure, and typically are a demonstration that the average probability of error decreases exponentially with code "block" length. Such random coding bounds only imply that codes which perform better than the average exist and do not indicate how good codes can be found.

Clearly, the coding bounds are not helpful if we need to evaluate the performance of a specific code, and the performance of particular codes is often difficult to calculate. An important generalization of the Hamming codes for multiple-error correction is the class of codes called BCH (Bose, Chaudhuri, Hocquenghem) codes. Figure 3.11 compares the bit error probability (P_b) of several (n,k) BCH codes as a function of $E_b/\mathcal{N}_0$ for binary PSK with hard decisions where the channel bit error probability $P = Q[\sqrt{2kE_b/n\mathcal{N}_0}]$. Note that the (7,4) code shown is a Hamming code. It is evident from this figure that the BCH codes can significantly improve performance in comparison to uncoded binary PSK (dashed line). It is noted that P_b in Fig. 3.11 is not exact but is an upper bound on the bit error probability of the codes (Clark and Cain 1981). Despite this, considerable performance improvement is evident.

Some computer simulation results for binary convolutional codes with Viterbi decoding and hard decisions are presented in Fig. 3.12 as the constraint length is varied (Heller and Jacobs 1971). The improvement over uncoded transmission can be seen by comparison with the uncoded curve in Fig. 3.11. Reduction in bit error probability as the constraint length is increased is clear from Fig. 3.12. By referring back to Example 3.4.1, the reader can see that the Viterbi algorithm must search to some (as yet unspecified) depth in the trellis before the best path is found, or more generally, before all retained paths spring from the same first step. When implementing the Viterbi algorithm, this depth must be selected. If this depth is chosen too small, there is ambiguity in the decoding decision [see, e.g., Fig. 3.10c where two paths have distance 2 and a different first step]. If this depth is chosen large, excessive storage is required. This "search depth" is generally selected to be several times (say, 5–10 times) the constraint length of the code as a compromise. The simulations for Fig. 3.12 used a search depth of 32. Figure 3.13 presents comparative performance results as the search depth or decoder memory, denoted by λ, is varied for a $v = 5$, rate-$\frac{1}{2}$ convolutional code. The three curves labeled $Q = 2$ are for the hard decision case. The $\lambda = 32$ choice yields a slight improvement over the $\lambda = 16$ depth.

Fig. 3.11 Bit error probability for BCH codes using binary PSK modulation and hard decisions. From G. C. Clark, Jr., and J. B. Cain, *Error-Correction Coding for Digital Communications,* New York: Plenum Press, 1981. © 1981 Plenum Press

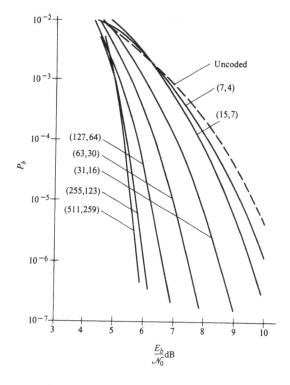

Fig. 3.12 Bit error probability for convolutional codes using PSK and hard decisions. From J. A. Heller and I. M. Jacobs, "Viterbi Decoding for Satellite and Space Communications," *IEEE Trans. Commun. Technol.,* © 1971 IEEE

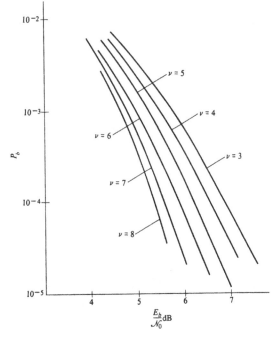

Fig. 3.13 Convolutional code
performance as a function of
decoder memory or search
depth (λ). From J. A. Heller
and I. M. Jacobs, "Viterbi
Decoding for Satellite and
Space Communications," *IEEE
Trans. Commun. Technol.,* ©
1971 IEEE

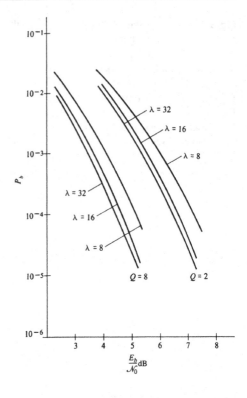

   Thus far in this chapter we have limited consideration to hard decision decoding, where
the output of the demodulator is simply a binary sequence. In this case, optimum (also
maximum a posteriori and minimum probability of error) decoding simplifies to calculating the appropriate Hamming distances. Hard decision decoding corresponds to a
two-level quantization of the decoder input. We know that we could better implement the
optimum decision rule for an AWGN channel if the demodulator output were unquantized, often called infinite quantization, or at least quantized more finely than to two
levels. The reader should review optimum receivers for unquantized outputs in Sect. 2.4
if this notion seems fuzzy. When the number of levels (Q) at the demodulator output is
greater than 2 for binary transmitted symbols, the system is said to be using soft decision
demodulation or soft decision decoding. Soft decision decoding can be used with either
block or convolutional codes, although it is much simpler to use with convolutional codes.
Figure 3.13 contains performance results for eight-level ($Q = 8$) soft decision decoding
as well as for hard decision decoding. The soft decisions have a clear performance advantage. When using soft decisions, there is always a trade-off between receiver complexity
and performance, both of which increase with the number of output quantization levels.
Figure 3.14a–d present the performance of rate $-\frac{1}{3}, -\frac{1}{2}, -\frac{2}{3}$, and $-\frac{3}{4}$ convolutional codes
for various constraint lengths (v) when used with binary PSK over an AWGN channel and

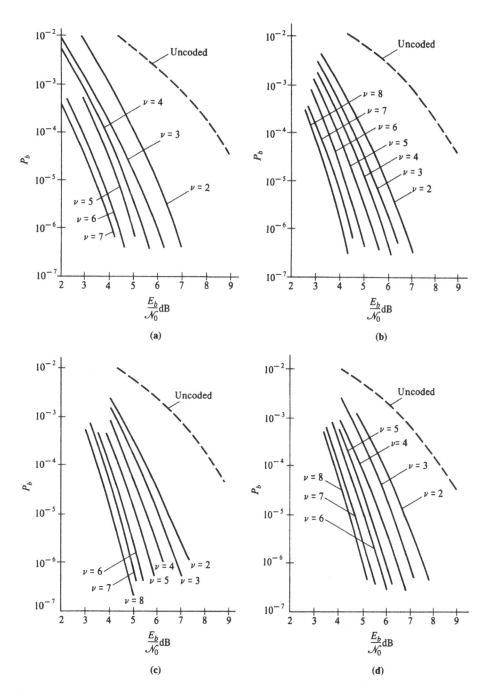

Fig. 3.14 Convolutional code performance with PSK modulation over an AWGN channel with soft decision Viterbi decoding: **a** $R_C = \frac{1}{3}$; **b** $R_C = \frac{1}{2}$; **c** $R_C = \frac{2}{3}$; **d** $R_C = \frac{3}{4}$. From G. C. Clark, Jr., and J. B. Cain, *Error-Correction Coding for Digital Communications,* New York: Plenum Press, 1981. © 1981 Plenum Press

infinitely quantized soft decision Viterbi decoding (Clark and Cain 1981). The lower rate codes and the longer constraint lengths for a given rate code provide better performance.

The question often arises as to which type of code, a block code or a convolutional code, is better for a particular application. Whether block codes or convolutional codes are better in general is a point of continuing debate among coding theorists. McEliece (1977) states that the BCH (block) codes are to be preferred for use over a BSC and other symmetric DMCs, since these are the channels the BCH codes were designed for, whereas for the AWGN channel, soft decision Viterbi decoded convolutional codes are preferable. This is primarily because of the ease with which soft decisions can be combined with Viterbi decoding. He also states that convolutional codes show up in more practical applications than block codes because the AWGN channel occurs more often in practical communication situations. Obviously, we cannot give a clear answer here as to whether one should use a block code or a convolutional code. The choice depends on the particular application of interest.

3.7   Summary

We have touched on a few of the most important concepts involved in forward error correction and error detection. Both block codes and convolutional codes have played a fundamental role in many communication systems and their analysis and design continues to be an area of ongoing research interest. Automatic repeat-request systems implement strategies that allow the powerful error detection capabilities of codes to be used to great advantage and new ARQ strategies will certainly be developed. The proper integration of error control codes into practical communication systems has lagged the development of the codes and coding theory, with the result that many existing communication systems do not perform as well as they could. For the application of coding to spread, it is important that undergraduate engineers, as a minimum, grasp the material in this chapter.

Problems

3.1   With reference to Fig. 3.1, assume that the rate of the binary message sequence is R_s = 20,000 bits/s. If the rate of the channel code is $R_c = \frac{2}{3}$, what is the rate transmitted over the discrete channel?

3.2   Which of the following generator matrices are systematic? What is the rate of the corresponding codes?

(a) $\mathbf{G}_1 = \begin{bmatrix} 1\ 0\ 0\ 0\ 1\ 1\ 0 \\ 0\ 1\ 0\ 0\ 0\ 1\ 1 \\ 0\ 0\ 1\ 0\ 1\ 1\ 1 \\ 0\ 0\ 0\ 1\ 1\ 0\ 1 \end{bmatrix}$

(b) $\mathbf{G}_2 = \begin{bmatrix} 1\ 1\ 1\ 0\ 0\ 0\ 0 \\ 0\ 0\ 1\ 1\ 0\ 1\ 0 \\ 0\ 1\ 1\ 0\ 1\ 1\ 0 \\ 0\ 1\ 0\ 0\ 0\ 1\ 1 \end{bmatrix}$

(c) $\mathbf{G}_3 = \begin{bmatrix} 1\ 0\ 0\ 1\ 1 \\ 0\ 1\ 0\ 0\ 1 \\ 0\ 0\ 1\ 0\ 1 \end{bmatrix}$

3.3   Assume that the codes in Problem 3.2 are to be used over a memoryless channel and write each generator matrix in the form of Eq. (3.2.2). Construct the corresponding parity check matrices.

3.4   For each of the generator matrices in Problem 3.2, construct codeword tables analogous to Table 3.1. Compare the codes generated by $\mathbf{G}_1$ and $\mathbf{G}_2$ with the (7,4) Hamming code in Table 3.1. This may be easier to do symbolically using Eq. (3.2.3).

3.5   Find the generator matrix for the parity check matrix

$$\mathbf{H} = \begin{bmatrix} 1\ 1\ 0\ 1\ 0\ 0 \\ 0\ 1\ 1\ 0\ 1\ 0 \\ 1\ 1\ 1\ 0\ 0\ 1 \end{bmatrix}.$$

If the message $\mathbf{v} = (1\ 0\ 1)$ is to be transmitted, find the corresponding codeword and show that its syndrome is $\mathbf{0}$.

3.6   For the (7,4) Hamming code in Example 3.2.1, decode the received sequence $\mathbf{x} = (0\ 1\ 0\ 0\ 1\ 1\ 1)$.

3.7   Decoding by hand can be simplified and linear block decoding can be better understood by creating what is called the *standard array* for a code. For an (n,k) linear block code, the standard array is constructed as follows. The all-zero vector followed by the 2^k codewords are placed in a row. An error pattern is placed under the all-zero vector and a second row is generated by adding this error pattern to the codeword immediately above in the first row. To begin the third row, an error pattern that has not yet appeared anywhere in the table is selected and placed in the first column. This error pattern is then used to fill out the third row by adding this error pattern to the codeword at the top of each column. The process is continued until the table contains all $2^n - 2^k$ detectable error patterns below the codewords, and there will be 2^{n-k} rows corresponding to the correctable error patterns. The error patterns in the first column of each row should be chosen as the remaining vector of least Hamming weight. We now form the standard array for the code with generator matrix

$$\mathbf{G} = \begin{bmatrix} 1\ 1\ 1\ 0\ 0 \\ 0\ 0\ 1\ 1\ 0 \\ 1\ 1\ 1\ 1\ 1 \end{bmatrix}.$$

Table 3.6 Standard array for the (5,3) code in Problem 3.7

Syndrome	Coset leader							
00	00000	11100	11010	00110	11111	11001	00101	00011
01	00001	11101	11011	00111	11110	11000	00100	00010
10	01000	10100	10010	01110	10111	10001	01101	01011
11	10000	01100	01010	10110	01111	01001	10101	10011

The possible codewords are listed in the first row of Table 3.6 preceded by the zero vector. A minimum weight error pattern, in this case, 0 0 0 0 1, is placed in the first position of the second row and added to the codewords at the top of each column to fill out the row. Note that 0 0 0 1 0 and 0 0 1 0 0 appear in the second row, and hence the minimum weight vector chosen to begin row 3 is 0 1 0 0 0. The process is continued until there are $2^{n-k} = 2^{5-3} = 4$ rows, as given in Table 3.6. The standard array contains all 24 detectable error patterns. To use this table for decoding, we need to involve the syndrome. It is interesting that all binary sequences in a row have the same syndrome (see Problem 3.10). Thus we need only calculate the syndrome for each word in the first column, which is called the *coset leader*. These are shown in Table 3.6. Decoding using the standard array therefore consists of the following steps: (1) The syndrome of the received vector is calculated; (2) the coset leader corresponding to the calculated syndrome is assumed to be the error pattern; (3) the error pattern is added to the received vector to get the decoded sequence. Given the received vector $\mathbf{x} = (1\,0\,1\,1\,1)$, use the standard array to decode this sequence.

3.8   Generate the standard array for the code in Problem 3.7 using 0 0 0 1 0 as a coset leader. Is this a good code? This illustrates why the coset leaders are called correctable error patterns.

3.9   Generate the standard array for the (7,4) Hamming code and use it to decode the received vector in Example 3.2.1.

3.10 Prove that all binary sequences in a row of the standard array have the same syndrome.

3.11 Let A_i, $i = 0, 1, 2, \ldots, n$, be the number of code vectors with Hamming weight i for a particular code. The A_i are called the *weight distribution* of the code and are very useful for probability of error calculations. Find the weight distribution of the (7,4) Hamming code in Table 3.1.

3.12 Find the weight distribution for the code with the generator matrix

$$\mathbf{G} = \begin{bmatrix} 0\,1\,1\,1\,0\,0 \\ 1\,0\,1\,0\,1\,0 \\ 1\,1\,0\,0\,0\,1 \end{bmatrix}.$$

3.13 There are $2^k - 1$ undetectable error patterns for an (n,k) linear block code that occur when an error pattern is the same as a nonzero codeword. Write the probability of an undetected error, $P_{ud}(\varepsilon)$, in terms of the weight distribution of the (7,4) Hamming code and the BSC bit error probability p. Evaluate $P_{ud}(\varepsilon)$ for $p = 10^{-1}, 10^{-2}, 10^{-3}, 10^{-4}$ and 10^{-5}.

3.14 Evaluate Eq. (3.2.21) for $n = 7$, $t = 1$, and $p = 10^{-1}, 10^{-2}, 10^{-3}$, and 10^{-4}.

3.15 Use Eq. (3.3.4) and $g(D) = D^3 + D + 1$ to generate a (7,4) cyclic code. Compare to the Hamming code in Table 3.1.

3.16

(a) Write the polynomial corresponding to the code vector cyclically shifted by i places,

$$\mathbf{w}^{(i)} = (w_{n-i+1} \ldots w_n w_1 \ldots w_{n-i}).$$

(b) Form $D^i w(D)$ and show that $D^i w(D) = f(D)(D^n + 1) + w^{(i)}(D)$, where $w^{(i)}(D)$ is the polynomial in part (a). Specify $f(D)$.

3.17 For the (7,3) cyclic code with generator polynomial $g(D) = (1 + D)(1 + D + D^3)$, find the generator matrix and the corresponding parity check matrix in systematic form.

3.18 Repeat Problem 3.17 for the (15,11) cyclic code with generator polynomial $g(D) = 1 + D + D^4$.

3.19 Draw encoder and decoder circuits for the (7,3) cyclic code with generator polynomial $g(D) = (1 + D)(1 + D + D^3)$.

3.20

(a) Using the encoding circuit in Fig. 3.3, find the codeword for $\mathbf{v} = (1\ 1\ 0\ 1)$. Show the shift register contents for all shifts of the circuit,

(b) If the error pattern, $\mathbf{e} = (0\ 0\ 1\ 0\ 0\ 0\ 0)$, is added to the code vector of part (a), use Fig. 3.4 to find the syndrome and decode.

3.21 An error burst of i bits at the beginning of the codeword and an error burst of j bits at the end of a codeword is called an *end-around* burst of length $i + j$. Show that an (n,k) cyclic code can detect end-around bursts of length $n - k$ by showing that $s(D) \neq 0$.

3.22 Show that the fraction of undetectable bursts of length $n - k + 1$ is $2^{-(n-k-1)}$.

3.23 What can you say about the burst error-correcting capability of the cyclic (7,4) code generated by $g(D) = 1 + D + D^3$? The (15,11) code generated by $g(D) = 1 + D + D^4$? See also Problem 3.22.

3.24 A polynomial $x(D)$ of degree m is said to be irreducible if $x(D)$ cannot be divided by any polynomial of degree m or less ($m_k \neq 0$). Show that $D^2, D^2 + 1$, and $D^2 + D$ are not irreducible but $1 + D + D^2$ is irreducible.

3.25 A primitive polynomial $x(D)$ can be defined as an irreducible polynomial of degree m, where the smallest integer n for which $x(D)$ divides $D^n + 1$ is $n = 2^m - 1$. Show that $1 + D + D^3$ divides $1 + D^7$ and that $1 + D + D^4$ divides $1 + D^{15}$. To show that these polynomials are primitive would require checking $1 + D^n$ for all $1 \leq n \leq 2^m - 1$ in each case (Lin and Costello 1983). A binary cyclic code with block length $n = 2^m - 1$ is called a primitive cyclic code (Blahut 1983).

3.26 If the input sequence to the (2,1) binary convolutional encoder in Fig. 3.6 is $\mathbf{v} =$ (1 1 0 0 1 0 1 0 0 $\cdots$), find the encoder output sequence $\mathbf{w}$ from this figure.

3.27 Repeat Problem 3.26 using Eq. (3.4.5).

3.28 The rate-$\frac{1}{3}$ convolutional encoder in the figure below is a modification of the rate-$\frac{1}{2}$ convolutional encoder in Fig. 3.6, obtained by feeding the input bit directly to the output. For the input sequence

$$\mathbf{v} = \left(1\ 0\ 1\ 1\ 0\ 1\ 0\ 0 \cdots \right),$$

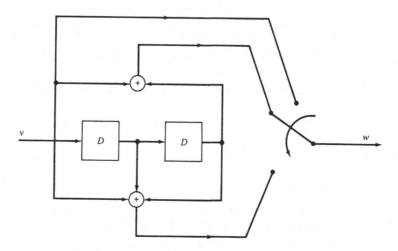

find the encoder output sequence $\mathbf{w}$. Sketch the alternative block diagram for this encoder analogous to that shown in Fig. 3.7.

3.29 Specify the generator sequences for the encoder in the above figure and find the output sequence $\mathbf{w}$ for $\mathbf{v} = \left(1\ 0\ 1\ 1\ 0\ 1\ 0\ 0 \cdots \right)$ using Eq. (3.4.3).

3.30 For the convolutional encoder in the above figure, write the generator matrix and find the output $\mathbf{w}$ for $\mathbf{v} = \left(1\ 0\ 1\ 1\ 0\ 1\ 0\ 0 \cdots \right)$ using Eq. (3.4.5).

3.31 Specify the constraint length of the convolutional encoder in the above figure and draw and label the code tree for this encoder to depth 5.

3.32 Starting with the code tree in Problem 3.31, draw the trellis for the convolutional encoder in the above figure.

3.33 The movement through a trellis as demonstrated in Fig. 3.9 is simply a set of transitions between a finite number of states, where all states may not be reachable from any other state. A convolutional encoder thus has an interpretation as what is called a *finite state machine*. A finite state machine can be represented by a *state diagram* that shows the states as circles and the possible state transitions. State diagrams play an important role in the analysis and understanding of convolutional encoders. The state diagram for the trellis in the figure below shows, where the states are shown as appropriately labeled circles and the state transitions by directed arrows, with the message bit above the arrow and the encoder output bits in parentheses below the arrows. Convince yourself of the validity of this state diagram and use it to find the encoder output **w** for the input message sequence $\mathbf{v} = \left(1\,0\,1\,1\,0\,1\,0\,0\,\cdots \right)$.

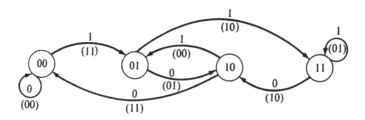

3.34 Draw the state diagram (see Problem 3.33) for the trellis in Problem 3.32.

3.35 The figure below shows the convolutional encoder for a rate-$\frac{2}{3}$ code. The sequence of message bits is now used two at a time to generate encoder outputs by alternately applying them to the upper and lower inputs. For the input sequence $\mathbf{v} = \left(1\,1\,0\,0\,1\,0\,1\,1\,0\,0\,1\,0\,0\,0\,\cdots \right)$, find the output sequence for this encoder. The constraint length for such a code is sometimes said to be $v + k$.

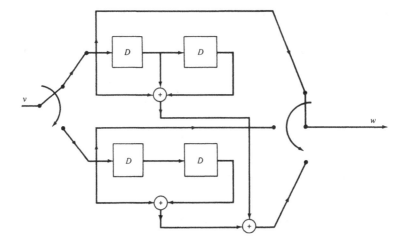

3.36 Draw the tree corresponding to the encoder in the figure above.

3.37 Use the results of Problem 3.36 to draw the trellis and state diagram (see Problem 3.33 for the definition) for the convolutional encoder (see Problem 3.33 figure).

3.38 The rate-$\frac{1}{2}$ binary convolutional code in Fig. 3.6 with the trellis in Fig. 3.9 is used and the transmitted codeword is the all-zero sequence $\mathbf{w} = \left(0\,0\,0\,0\,0\,0\,0\,0 \cdots \right)$. The received vector is

$$\mathbf{x} = \left(1\,1\,0\,0\,0\,0\,0\,0 \cdots \right).$$

Use the Viterbi decoder to decode the received sequence.

3.39 Repeat Problem 3.38 if the received sequence is

$$\mathbf{x} = \left(1\,1\,1\,0\,0\,0\,0\,0 \cdots \right).$$

3.40 For the rate-$\frac{1}{3}$ convolutional encoder (see Problem 3.28 figure), the all-zero codeword is transmitted and the received vector is

$$\mathbf{x} = \left(1\,1\,0\,1\,0\,0\,0\,0\,0 \cdots \right).$$

Use Viterbi decoding to decode this received sequence.

3.41 In this chapter we have discussed only linear codes. Further, in our studies of Viterbi decoding, we have always assumed that the transmitted codeword is all zeros. It turns out that in all analyses of Viterbi decoder performance, we can assume without loss of generality that the all-zeros sequence is transmitted. Why?

Hint: See McEliece (1977).

3.42 Although exact calculation of the exact error probability for various codes as shown in Figs. 3.11, 3.12, 3.13 and 3.14 is difficult, in general, it is often possible to determine bounds on the error probability that are sufficient to evaluate the code performance. For example, consider the use of an (n,k) block code to represent one of M messages for transmission with BPSK modulation over an additive white Gaussian noise channel with a two-sided power spectral density of $\mathcal{N}_0/2$ W/Hz . Let the energy per symbol be E_s so that the energy per information bit is $E_b = E_s/k$. If we define E_c to be the energy per codeword bit and $R_c = k/n$, then also $E_b = E_c/R_c$. The optimal coherent receiver observes the n demodulator outputs given by $r_j = \pm E_c + n_j$, $j = 1, 2, \ldots, n$, when n_j is the sampled additive white Gaussian noise, and forms the test statistic

$$Z_i = \sum_{j=1}^{n} \left(2c_{ij} - 1 \right) z_j, \quad i = 1, 2, \ldots, M,$$

where c_{ij} is the jth bit of the ith codeword.

(a) Assume that the all-zeros codeword was transmitted, with $\mathbf{c}_1 = \begin{bmatrix} c_{11} & c_{12} & \ldots & c_{1n} \end{bmatrix} = \begin{bmatrix} 0\,0\,\ldots\,0 \end{bmatrix}$ find $E[Z_1|\mathbf{c}_1]$, $E[Z_i|\mathbf{c}_1]$, $i = 2, 3, \ldots, M$, and var $[Z_i|\mathbf{c}]$ for all i.

(b) Show that

$$P[Z_j > Z_1|\mathbf{c}_1] = \frac{1}{2} - \mathrm{erf}\left[\sqrt{\frac{2E_s w_j}{n \mathcal{N}_0}}\right]$$

For all J = 1 where W_J is the Hamming weight of the Jth codeword

(c) Since each of the Z_i's is equally likely, use the union bound to obtain

$$P_e \leq \sum_{j=2}^{M}\left[\frac{1}{2} - \mathrm{erf}\left(\frac{2E_b}{\mathcal{N}_0}w_j R_c\right)\right]^{1/2}$$

3.43 Note that the error probability in part (c) of Problem 3.42 requires knowledge of the code weight distribution.

(a) Show that another bound on error probability is given by

$$P_e \leq (M-1)\left[\frac{1}{2} - \mathrm{erf}\left(\frac{2E_b}{N_0}R_c d_{\min}\right)^{1/2}\right]$$

where $d_{\min}$ is the code minimum distance.

(b) A quantity called the *coding gain* is often defined, which is the difference in SNR required without and with coding to achieve a given error probability. For BPSK without coding,

$$P_e = \frac{1}{2} - \mathrm{erf}\left[\sqrt{\frac{2E_b}{\mathcal{N}_0}}\right] < \frac{1}{2}e^{-2E_b/\mathcal{N}_0},$$

where the last inequality is a commonly used bound (Proakis 1989). Show that the result from part (a) with coding can be bounded as

$$P_e < \frac{M}{2}e^{(-2E_b/\mathcal{N}_0)R_c d_{\min}}$$

(c) By comparing the two bounds in part (b), show that the coding gain for an (n,k) block code is given by

$$10\log_{10}\left(\frac{R_c d_{\min}}{2} - \frac{k\mathcal{N}_0}{2E_b}\ln 2\right)\mathrm{dB}.$$

Coded Modulation

4

4.1   Introduction

For the digital communications systems discussed thus far in the book, we have assumed that coding (operations on the binary data stream) and modulation are independent in the sense that the modulator simply accepts the binary sequence presented to it and maps a set of one or more bits, in a one-to-one fashion, onto a waveform appropriate for transmission over the channel. This is the most widely accepted approach to digital communications today, and with reference to the signal space diagrams in Chap. 2, it implies that only two parameters can be adjusted to improve communication system performance: transmitted power and channel bandwidth. Specifically, for a fixed data rate, the symbol error probability can be reduced by increasing the transmitted power, or for a fixed error probability, the data rate can be increased if the available bandwidth is increased. A second implication of this separation of coding and modulation is that if channel coding (as described in Chap. 3) is used, a higher data rate results, which necessitates a wider bandwidth. Thus it seems that whatever we do to improve digital communication system performance either increases the transmitted power or the required bandwidth.

Fortunately, since transmitted power and channel bandwidth are precious commodities today, there is a way out of this dilemma. The basic concept involved is that of observing an entire sequence or block of data before making a decision at the receiver, rather than operating on a symbol-by-symbol basis. To be more exact, consider a sequence of symbols of length N. If each symbol is a point in a two-dimensional signal constellation as in Chap. 2, the sequence of symbols is a point in $2N$-dimensional space. The idea is thus to consider the Euclidean distance between *sequences* in this $2N$-dimensional space, and the signal design problem is to select the allowable *sequences* such that they are as far apart as possible. As long as the message bit rate, or bit rate to be transmitted, is small enough that some of the points in the two-dimensional signal constellation are unused, by

© The Author(s), under exclusive license to Springer Nature Switzerland AG 2023
J. D. Gibson, *Digital Communications*, Synthesis Lectures on Communications,
https://doi.org/10.1007/978-3-031-19588-4_4

letting N get large, the error probability can be reduced. Coding becomes intertwined with modulation, since channel codes are employed to specify the set of allowable transmitted sequences in the $2N$-dimensional space. This combined modulation and coding approach fits within the theory provided 20 years ago by Wozencraft and Jacobs (1965), and when convolutional codes (in contrast to block codes) are employed, the approach is called *trellis-coded modulation* (TCM).

Spectrally efficient constant envelope modulation techniques are related to trellis-coded modulation methods, although they are not the same, and there is some debate as to whether spectrally efficient constant envelope modulation falls within the theoretical framework described by Wozencraft and Jacobs (1965). However, the combination of coding and modulation is evident in these techniques. The requirement for a constant envelope in the transmitted waveform follows from the fact that nonlinearities present in some communications channels, such as satellite channels, can distort as well as translate envelope variations into other types of distortion. Holding the envelope constant naturally implies that we are using either phase or frequency modulation. We know that FM and PM can require relatively large bandwidths, primarily because of rapid changes in transmitted phase. Spectrally efficient constant envelope methods only allow "smooth" changes in the phase, thus reducing bandwidth requirements, and accommodate an input message that requires a large total phase change by spreading this phase change out over several symbol intervals. Thus, to recover the total phase change corresponding to a particular message in a particular symbol interval, the receiver must examine a sequence of phases and compare it to all allowable sequences of phase changes. Combined coding and modulation is thus again evident in these schemes, although not as explicitly as in TCM.

The common denominator in both types of coded modulation, TCM and constant envelope methods, is the necessity of comparing received sequences to allowable transmitted sequences and finding the "nearest" one. Therefore, the Viterbi algorithm and related trellis search algorithms are an integral part of the receivers for such modulations.[1] Thus, although we are getting improved performance without increasing transmitted power or channel bandwidth, we are not getting something for nothing. The price we are paying is in terms of increased receiver complexity or increased signal processing. However, this is not a major problem in many communication systems today because of the success of microelectronics and VLSI, which allow very complicated devices to be constructed in very small packages with lower power consumption. Furthermore, these devices may be quite inexpensive, particularly if there is a large commercial market, as is possible in telecommunications.

Section 4.2 illustrates the idea that comparing sequences can provide a performance improvement over symbol-by-symbol decoding/demodulation by considering binary FSK, which is a constant envelope technique. Spectrally efficient constant envelope modulation is then developed in Sect. 4.3, followed by a treatment of trellis-coded modulation in

---

[1] As a result, the reader should probably be familiar with Sect. 2.7 and Chap. 3 before proceeding here.

Sect. 4.4. The summary in Sect. 4.5 tries to place these techniques in perspective and to whet the reader's appetite for a more detailed study of the ideas involved.

4.2   Coherent Binary FSK

In this section we consider a particular example of constant envelope modulation due to de Buda (1972), which is included to demonstrate explicitly that observing more than one symbol interval at the receiver can improve communication system performance. This example is not crucial to the developments in subsequent sections but is provided for motivational purposes. Following de Buda (1972), we examine coherent binary frequency shift keying (FSK), and we represent the transmitted waveform at any time instant by its pre-envelope

$$s(t) = e^{j\{(\omega_1 + \omega_2)t/2 + \phi(t)\}} \tag{4.2.1}$$

where ω_1 corresponds to a 1 being transmitted, ω_2 corresponds to a 0, and $\phi(t)$ is varied to yield the desired transmitted signal. We use the analytic signal representation because it is simple to manipulate. The actual transmitted waveform is the real part of $s(t)$, and the pre-envelope representation only requires the assumption that the signal is bandpass [see Problem 4.3 and Rice (1982)].

Letting $s_0(t)$ denote the pre-envelope when a 0 is transmitted, then we find that

$$s_0(t) = e^{j[\omega_2 t + \phi(0)]}, \quad 0 \le t \le T, \tag{4.2.2}$$

which upon comparing Eqs. (4.2.1) and (4.2.2) implies that

$$\phi(t) - \phi(0) = \frac{\omega_2 - \omega_1}{2}t. \tag{4.2.3}$$

Thus, at time $t = T$,

$$\begin{aligned} \phi(T) - \phi(0) &= \frac{\omega_2 - \omega_1}{2}T \\ &= \pi(f_2 - f_1)T, \end{aligned} \tag{4.2.4}$$

so if $\omega_2 > \omega_1$, the phase increases by $\pi(f_2 - f_1)T$ in T seconds. Defining $s_1(t)$ as the pre-envelope when a 1 is transmitted, we obtain

$$s_1(t) = e^{j[\omega_1 t + \phi(0)]}, \quad 0 \le t \le T, \tag{4.2.5}$$

and it is straightforward to show that the phase decreases by $\pi(f_2 - f_1)T$ T when a 1 is sent. It is common to define a parameter h called the modulation *index* by

$$h = (f_2 - f_1)T. \tag{4.2.6}$$

Fig. 4.1 Possible phase
differences $\phi(t) - \phi(0)$ as a
function of time

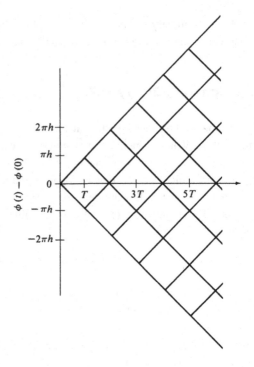

If at time $t = (j + k)T$, j 0's and k 1's have been sent, the phase is given by

$$\phi[(j + k)T] - \phi[0] = (j - k)h\pi \tag{4.2.7}$$

The possible phase differences $\phi(t) - \phi(0)$ can be sketched as a function of time as shown
in Fig. 4.1.

It is of great interest to find the value of the modulation index h that minimizes the
probability of error. If we limit consideration to the AWGN channel, we know from
Chap. 2 that we need to select h to maximize the distance between $s_0(t)$ and $s_1(t)$. Thus
we need to maximize

$$d^2(s_0, s_1) = \int_0^T |s_0(t) - s_1(t)|^2 dt \tag{4.2.8}$$

with respect to h. It can be shown by straightforward manipulations that

$$|s_0(t) - s_1(t)| = 2 \sin \frac{\pi h t}{T}, \tag{4.2.9}$$

so Eq. (4.2.8) yields

$$d^2(s_0, s_1) = 4 \int\limits_0^T \sin^2 \frac{\pi h t}{T} dt = 2T \left[1 - \frac{\sin(2\pi h/T)}{(2\pi h/T)} \right], \qquad (4.2.10)$$

the second term of which has a minimum of -0.217 when $\tan 2\pi h = 2\pi h$, or when $h = 0.714$. Therefore, h_{opt} seems to be 0.714.

The reason that we say "seems to be" in this last sentence is that in deriving this value for h, we have made an incorrect implicit assumption concerning the integration limits in Eq. (4.2.8). Although integrating over one symbol interval seems intuitive, we show in what follows that using a different value of h and integrating over two symbol intervals yields improved performance. We begin with Eq. (4.2.1) and write for $0 \le t \le T$,

$$s(t) = e^{j\{(\omega_1 + \omega_2)t/2 + \phi(t) - \phi(0) + \phi(0)\}}$$

$$= e^{j\{(\omega_1 + \omega_2)t/2 \pm \pi h t/T + \phi(0)\}}, \qquad (4.2.11)$$

where we have substituted for $\phi(t) - \phi(0)$, with the $+$ sign for a 0 and the $-$ sign for a 1. For $-T \le t \le 0$, $s(t)$ is the same as in Eq. (4.2.11) except that the signs on the $\phi(t) - \phi(0)$ term are reversed. To recover the phase (message) sequence, we multiply $s(t)$ by $e^{-j(\omega_1 + \omega_2)t/2}$ and take the real part, so that

$$r(t) \triangleq \text{Re}\left[s(t) e^{-j(\omega_1 + \omega_2)t/2} \right]$$

$$= \begin{cases} \cos\left[\pm\left(\frac{\pi h t}{T}\right) + \phi(0)\right], & 0 \le t \le T, \\ \cos\left[\mp\left(\frac{\pi h t}{T}\right) + \phi(0)\right], & -T \le t \le 0. \end{cases} \qquad (4.2.12)$$

Now, according to Eq. (4.2.7) with $h = 0.5$, the phase difference is either 0 or π at even multiples of T; hence, considering 0 an even multiple of T, we let $\phi(0) = 0$ or π. Equation (4.2.12) thus becomes

$$r(t) = \begin{cases} \cos\left[\pm\frac{\pi t}{2T}\right], & 0 \le t \le T, \ \phi(0) = 0 \\ -\cos\left[\pm\frac{\pi t}{2T}\right], & 0 \le t \le T, \ \phi(0) = \pi \\ \cos\left[\mp\frac{\pi t}{2T}\right], & -T \le t \le 0, \ \phi(0) = 0 \\ -\cos\left[\mp\frac{\pi t}{2T}\right], & -T \le t \le 0, \ \phi(0) = \pi, \end{cases}$$

which, since $\cos[\pm\pi t/2T] = \cos[\mp\pi t/2T]$, collapses to

$$r(t) = \begin{cases} \cos\frac{\pi t}{2T}, & \phi(0) = 0 \\ -\cos\frac{\pi t}{2T}, & \phi(0) = \pi \end{cases} \qquad (4.2.13)$$

for $-T \le t \le T$. From this equation it is evident that we can decide what phase was transmitted at $t = 0$, either $\phi(0) = 0$ or $\phi(0) = \pi$, by observing $r(t)$ over $-T \le t \le T$. Thus $r(t)$ consists of binary antipodal signals over $-T \le t \le T$, and has an error probability in terms of signal-to-noise ratio (SNR) given by [see Eq. (2.5.9)]

$$P_{\text{opt}}[\mathcal{E}] = \frac{1}{2} - \text{erf}[\sqrt{\text{SNR}}], \qquad (4.2.14)$$

which is 3 dB better than coherently orthogonal (over $0 \le t \le T$) FSK, which has [see Eq. (2.5.10)]

$$P_{\text{FSK}}[\mathcal{E}] = \frac{1}{2} - \text{erf}\left[\sqrt{\frac{\text{SNR}}{2}}\right] \qquad (4.2.15)$$

Observing more than one symbol interval thus provides improved system performance. The following section delves further into these constant envelope modulation methods.

4.3   Constant Envelope Modulation

We begin by considering a constant envelope, phase-varying sinusoid of the form

$$s(t) = \sqrt{\frac{2E}{T}} \cos[\omega_c t + \phi(t)], \qquad (4.3.1)$$

where T is the symbol interval, E the symbol energy, ω_c the carrier frequency, and $\phi(t)$ the time-varying phase. We are presented with a sequence of underlying phase changes, corresponding to the message sequence to be transmitted, that is represented by $\{a_i, i = \ldots, -2, -1, 0, 1, 2, \ldots\}$, where each a_i is a member of the M-ary alphabet $\pm 2\pi h, \pm 3(2\pi h), \ldots, \pm(M-1)(2\pi h)$ for M even, or $0, \pm 2(2\pi h), \ldots, \pm(M-1)(2\pi h)$ for M odd. The constant parameter h is called the *modulation index,* and the phase $\phi(t)$ in Eq. (4.3.1) changes in accordance with the data sequence $\{a_i\}$ through the relationship

$$\phi(t) = \sum_{i=-\infty}^{\infty} a_i f(t - iT), \quad -\infty < t < \infty, \qquad (4.3.2)$$

where $f(t)$ is a phase smoothing function to be chosen. If $\phi(t)$ is a continuous function of time, we obtain a spectrally efficient form of constant envelope modulation called *continuous phase modulation* (CPM). The function $f(t)$ is called a *phase pulse* and is sometimes defined in terms of a frequency pulse $g(t)$ as

$$f(t) = \int_{-\infty}^{t} g(\tau)d\tau, \quad -\infty < t < \infty, \qquad (4.3.3)$$

Let a_n be the phase change associated with the interval $[(n-1)T, nT]$, where the change occurs just after time $(n-1)T$. The smoothing function $f(t)$ causes this change to be spread over the next L symbol intervals. The phase pulse $f(t)$ can be any function that is zero before some (possibly negative) value of t and is $\frac{1}{2}$ for some sufficiently large value of t. Thus, if we consider two possible amplitudes of the nth phase change, denoted by $a_n(1)$ and $a_n(2)$ such that $a_n(1) - a_n(2) = 2k(2\pi h)$, the carrier phases corresponding to $a_n(1)$ and $a_n(2)$ will eventually differ by $k(2\pi h)$ radians if all other phase changes $a_i, i \neq n$, in the sequences are the same. In terms of the frequency pulse $g(t)$, we have a *causal* CPM system if

$$g(t) \equiv 0, \quad \text{for } t < 0 \text{ and } t > LT$$
$$g(t) \not\equiv 0, \quad \text{for } 0 \le t \le LT$$

(4.3.4)

Figure 4.2 shows four possible phase smoothing functions $f(t)$ and their corresponding instantaneous frequency pulses $g(t)$ (Anderson et al. 1981). The $f(t)$ in Fig. 4.2a represents standard PSK, where there is a step change in the phase. The $f(t)$ in (b) causes the phase to change linearly over one interval, which, as is evident from $g(t)$, is the same as standard FSK. Upon letting $\alpha = t/T$, the third phase smoothing function can be written as

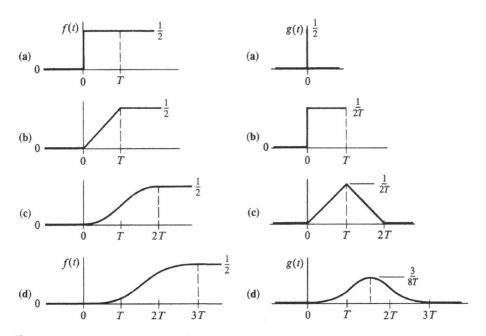

Fig. 4.2 Four possible phase smoothing functions and their corresponding frequency pulses. From J. B. Anderson, C.-E. W. Sundberg, T. Aulin, and N. Rydbeck, "Power-Bandwidth Performance of Smoothed Phase Modulation Codes," *IEEE Trans. Commun.*, © 1981 IEEE

$$f(\alpha) = \begin{cases} 0, & \alpha < 0 \\ \frac{\alpha^2}{4}, & 0 \le \alpha \le 1 \\ -\frac{\alpha^2}{4} + \alpha - \frac{1}{2}, & 1 \le \alpha \le 2 \\ \frac{1}{2}, & \alpha \ge 2 \end{cases} \qquad (4.3.5)$$

and the function in (d) is given by

$$f(\alpha) = \begin{cases} 0, & \alpha < 0 \\ \frac{\alpha^3}{12}, & 0 \le \alpha \le 1 \\ -\frac{\alpha^3}{6} + \frac{3\alpha^2}{4} - \frac{3\alpha}{4} + \frac{1}{4}, & 1 \le \alpha \le 2 \\ \frac{\alpha^3}{12} - \frac{3\alpha^2}{4} + \frac{9\alpha}{4} - \frac{7}{4}, & 2 \le \alpha \le 3 \\ \frac{1}{2}, & \alpha \ge 3 \end{cases} \qquad (4.3.6)$$

Note that all of the smoothing functions shown are causal, and that $L = 0, 1, 2$, and 3 for (a), (b), (c), and (d), respectively. Another popular choice for the smoothing function is the raised cosine class given by

$$f(\alpha) = \begin{cases} 0, & \alpha < 0 \\ \frac{\alpha}{2L} - \frac{\sin(2\pi\alpha/L)}{4\pi}, & 0 \le \alpha \le L \\ \frac{1}{2}, & \alpha \ge L \end{cases} \qquad (4.3.7)$$

where L is the number of intervals over which the phase is smoothed. In conjunction with CPM, $L = 1$ is called *full response* signaling and $L > 1$ is called *partial response* signaling. This nomenclature simply indicates that the phase change is ($L > 1$, partial response) or is not ($L = 1$, full response) spread over more than one symbol interval and is not necessarily the same as the baseband spectral shaping.

The most studied case of continuous-phase, constant envelope modulation is the scheme with $L = 1$, which is variously known by the label of *fast FSK* (FFSK), *minimum shift keying* (MSK), or as a special case of continuous-phase FSK (CPFSK). We investigate this modulation method in the following example.

Example 4.3.1 We are interested in studying MSK, which has $L = 1$ and $f(t)$ as in Fig. 4.2b expressible by

$$f(t) = \begin{cases} 0, & t \le 0 \\ \frac{t}{2T}, & 0 \le t \le T \\ \frac{1}{2}, & t \ge T \end{cases} \qquad (4.3.8)$$

With $h = \frac{1}{2}$ and $M = 2$, the input phase changes are $\pm\pi$, so that since $f(t) = \frac{1}{2}$ for $t \ge T$, the total phase change per input symbol is $\pm\pi/2$. From $g(t)$ in Fig. 4.2b, we see that what we are really doing is switching back and forth according to the input data sequence between two frequencies spaced $1/2\ T$ hertz apart. Since this spacing is one-half the spacing required for noncoherent orthogonality and used in most noncoherent FSK systems, MSK is often labeled fast FSK or FFSK. Furthermore, $1/2T$ is the minimum frequency spacing allowable for two FSK signals to be coherently orthogonal, which explains the title MSK.

If we assume that $\phi(0) = 0$, all possible phase trajectories starting at time $t = 0$ can be displayed by the tree diagram shown in Fig. 4.3. Suppose that the input binary message sequence consists of a series of $+1$ s or -1s and that we associate $+1$ with $a_i = +\pi$ and -1 with $a_i = -\pi$. Then for any given binary input message sequence, we can trace through the tree diagram in Fig. 4.3 and write the transmitted phase at any time instant. Thus, for the input binary sequence $-1, -1, +1, -1, -1, +1, +1, +1$, the corresponding transmitted phases at time instants nT, $n \geq 1$, are $-\pi/2, -\pi, -\pi/2, -\pi, -3\pi/2, -\pi, -\pi/2$, and 0.

Recovery of the input message sequence requires that the received phases be compared to all possible transmitted phase sequences using a tree search procedure, such as the Viterbi algorithm described in Sect. 2.7 and Chap. 3, and the "closest" path to the received sequence

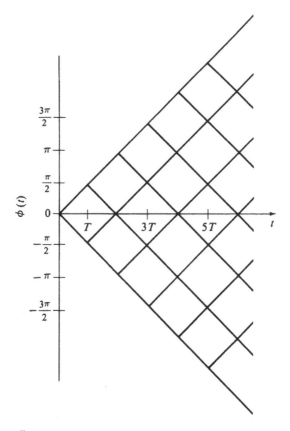

Fig. 4.3 MSK phase diagram

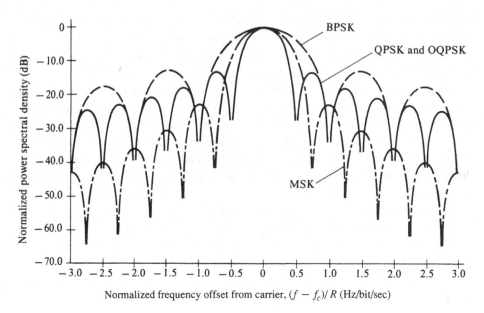

Fig. 4.4 Power spectral densities of BPSK, QPSK, and MSK. From F. Amoroso, "The Bandwidth of Digital Data Signals," *IEEE Commun. Mag.,* © 1980 IEEE

decoded as the input message sequence. Thus, although classical channel encoders are not used here, certainly, familiar decoders are being employed, and the boundary between coding and modulation is blurred. The complexity of the receiver over classical noncoherent FSK has clearly increased, but improved spectral efficiency is evident from Fig. 4.4, which compares the power spectral densities of binary PSK (BPSK), quadrature PSK (QPSK), and MSK. MSK has a narrower main lobe and much lower sidelobes than BPSK, and in comparison to QPSK, MSK has lower sidelobes but a wider main lobe. Thus, in terms of spectral efficiency, there is a trade-off between QPSK and MSK that depends on the particular application.

It is interesting to note that MSK can also be expressed and implemented as a shaped pulse version of offset or staggered QPSK. This connection is pursued in the problems.

Another variation on spectrally efficient constant envelope modulation is what is called *multi-h CPM*. The transmitted waveform for a multi-*h* CPM system has the same form as $s(t)$ in Eq. (4.3.1), but now $\phi(t)$ is given by

$$\phi(t) = \sum_{i=-\infty}^{\infty} a_i h_i f(t - iT), \quad -\infty < t < \infty, \qquad (4.3.9)$$

where the a_i and $f(t)$ are as before and the set $\{h_i, i = 1, 2, \ldots, K\}$ represents ordered modulation indices that are cyclically used proceeding as $h_1, h_2, \ldots, h_K, h_1, h_2, \ldots$ Just as

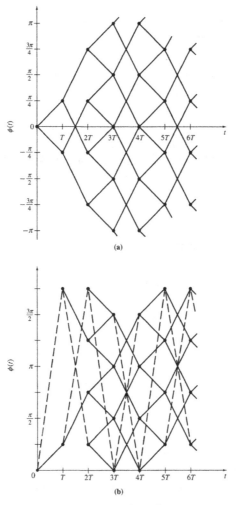

Fig. 4.5 **a** Tree and **b** trellis structures for a multi-$h\left\{\frac{1}{4}, \frac{2}{4}\right\} L = 1$ CPM system. From B. A. Mazur, and D. P. Taylor, "Demodulation and Carrier Synchronization of Multi-h Phase Codes," *IEEE Trans. Commun.,* © 1981 IEEE

in the constant-h case, any phase change is spread over L symbol intervals depending on $f(t)$, but h_i is changing from symbol to symbol. The h_i are chosen to be rational numbers. The multi-h CPM schemes produce phase trees just like the constant-h case. A phase tree for an $\{h_1, h_2\} = \left\{\frac{1}{4}, \frac{2}{4}\right\}$ CPM system with $L = 1$ (see Fig. 4.2b) is shown in Fig. 4.5a. Since phase is a modulo-2π quantity, the tree in part (a) can be drawn as a trellis as shown in Fig. 4.5b, where the dashed lines represent "wrap-around" state transitions (Mazur and Taylor 1981). Note that for a binary input sequence, the total transmitted phase changes by $\pm\pi/4$ radians when $h_1 = \frac{1}{4}$ is used and by $\pm\pi/2$ radians when $h_2 = \frac{2}{4}$ is used. The binary sequence $+1, -1, -1, -1, +1, +1, -1$, starting at $t = 0$, would thus yield transmitted phases at time nT of $\pi/4, -\pi/4, -\pi/2, -\pi, -3\pi/4, -\pi/4$, and $-\pi/2$.

Decoding of multi-h codes still consists of comparing the received phase sequence to all possible paths in the phase tree and selecting that possible path which is closest to the received sequence. However, calculating distances can be more difficult, since the transmitted signals need not be orthogonal, as they are, for instance, for MSK. This topic is beyond our intended development and is not pursued further here. The choice of sets of modulation indices must be done carefully, and different choices can yield quite different performances and spectra. Reduced spectral requirements for multi-h codes compared to MSK have been reported in the literature (Anderson and Lesh 1981; Ziemer and Peterson 1985).

4.4   Trellis-Coded Modulation

In this section we do not restrict ourselves to constant envelope modulations, and the use of coding as described in Chap. 3 will be seen to be more explicit. As noted in the introduction to this chapter and as in Sects. 4.2 and 4.3, the basic operational change over more familiar modulation methods is that the receiver bases its decision on an entire sequence of symbols rather than on a symbol-by-symbol calculation. The concept that allows such an operational technique to provide a performance improvement is described in Wozencraft and Jacobs (1965) and is explained geometrically as follows. Let each transmitted symbol be a point in a two-dimensional signal constellation, and let us assume that some of the points in the two-dimensional constellation are unused. To be more specific, let the number of available signal points in the two-dimensional constellation be A, but only $B < A$ of these points are actually used for transmitted signals. The ratio $B/A < 1$ remains fixed if the number of message bits/symbol is held fixed. Now we consider a sequence of two of these symbols. There will be A^2 available signal points, but only B^2 need to be used. If we consider a sequence of N symbols, these numbers become A^N and B^N, and the ratio of used signal points to unused signal points is $(B/A)^N$. Since $B/A < 1, (B/A)^N \to 0$ as $N \to \infty$; that is, the fraction of the possible transmitted symbols actually used becomes smaller as the length of the sequence increases. This implies that the distance between transmitted signals can be made larger with increasing N, and hence that the error probability can be reduced by increasing N. Not all choices of the B^N out of the A^N points are good ones (far apart), and coding combined with what is called "mapping by set partitioning" helps us select a good set of B^N points. Either block codes or convolutional codes can be employed; however, our development emphasizes convolutional codes, since they have received the most attention and seem to be better for coded modulation implementations. Note that channel coding alone cannot provide the available gain, and the mapping of the channel encoder output bits into symbols via the mapping by set partitioning procedure is critical to the success of the method.

To be more specific, we assume that we wish to transmit m bits/symbol, thus requiring a signal constellation of at least 2^m signal points. For coded modulation this signal set is expanded to 2^{m+1} (other expansions greater than 2^m will work) signals, so that in terms of our previous discussion, $B = 2^m$ and $A = 2^{m+1}$, yielding $B/A = \frac{1}{2}$. We now consider N symbol sequences and select the 2^{mN} best transmitted sequences out of the $2^{(m+1)N}$ possible transmitted sequences. In this manner we are able to keep the bit rate at m bits/symbol and still spread out the points in $2N$-dimensional space.

To expand the signal set from 2^m to 2^{m+1}, we use a rate $m/(m + 1)$ convolutional code with constraint length $m + 1$. Note that a block code will also do, but we stick with convolutional codes, for the reasons noted previously. The mapping by set partitioning procedure is then used to map these bits into channel signals. Since convolutional codes can be represented by a trellis structure, the rules for this mapping can be stated as follows (Ungerboeck 1982):

1. All parallel transitions in the trellis structure are allocated with the maximum possible Euclidean distance in the signal constellation.
2. All transitions diverging from or merging with a trellis state receive the next maximum possible Euclidean distance.

These rules guarantee that all single and multiple errors exceed the Euclidean distance of the uncoded m-bit/symbol constellation. Although this all seems mysterious at this point, the following example will help clarify the procedure.

Example 4.4.1 We consider the transmission of 2 bits/symbol using a rate-$\frac{2}{3}$, constraint length 3 convolutional code, and eight-phase PSK from Ungerboeck (1982). The partitioning of the eight-phase PSK signal constellation into subsets with increasing Euclidean distance is shown in Fig. 4.6. The trellis structure of the chosen convolutional code is given in Fig. 4.7, where the digits in the column "subset transmitted" indicate the transitions out of each node in top-to-bottom order. The parameter d_{free} is called the *maximum free distance* and is an indicator of the performance of the coded modulation scheme. The larger d_{free}, the better the performance. The convolutional encoder that generates the trellis in Fig. 4.7 is shown in Fig. 4.8. The subscript n denotes the time instant, x_n^1 and x_n^2 are the input bits, y_n^0, y_n^1, and y_n^2 are the output bits, s_n^1, s_n^2, and s_n^3 are bits representing the state, and a_n, is the transmitted symbol after mapping by set partitioning. For a review of convolutional encoding, see Chap. 3. Note that the particular diagram for this convolutional encoder as shown in Fig. 4.8 is called a minimal realization, since it is simpler than other realizations.

Using Figs. 4.7 and 4.8, we can find the encoder output and the transmitted symbol sequences for a given binary input sequence. For example, let the initial encoder state be $\left(s_0^3 s_0^2 s_0^1\right) = \left(0\ 0\ 0\right)$ and assume that the input sequence is 1 1 0 0 0 1 0 1 1 0 1 1. Table 4.1 shows the input, current state, coder output, path transition, transmitted symbol, and next

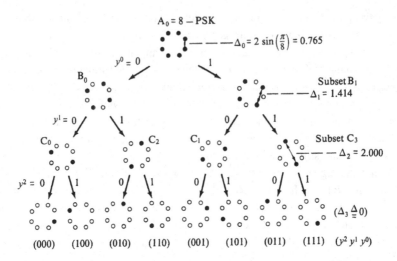

Fig. 4.6 Mapping by set partitioning of the eight-phase PSK signal set. From G. Ungerboeck, "Channel Coding with Multilevel/Phase Signals," *IEEE Trans. Inf. Theory,* © 1982 IEEE

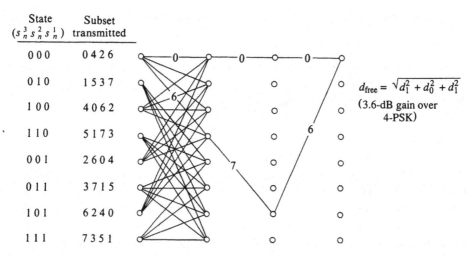

Fig. 4.7 Trellis diagram of the rate-2/3 constraint length 3 convolutional code indicating allowable phase transitions for 2 bits/symbol. From G. Ungerboeck, "Channel Coding with Multilevel/Phase Signals," *IEEE Trans. Inf. Theory,* © 1982 IEEE

state as a function of time n. By comparing the numbers in the transmitted signal column in Table 4.1 with the numbers in Fig. 4.6, the specific phases being transmitted can be determined.

To calculate the distance between two paths, we use the distances (d_i) shown in Fig. 4.6 and take the square root of the sum of the squared distances between each transmitted signal

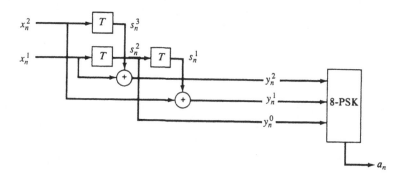

Fig. 4.8 Minimal realization of the convolutional encoder for the trellis in Fig. 4.7. From G. Unger-boeck, "Channel Coding with Multilevel/Phase Signals," *IEEE Trans. Inf. Theory,* © 1982 IEEE

Table 4.1 Coding and modulation process for Example 4.4.1

Time (n)	Input $(x_n^2 x_n^1)$	State $(s_n^3 s_n^2 s_n^1)$	Coder output $(y_n^1 y_n^2 y_n^3)$	Next state $(s_{n+1}^3 s_{n+1}^2 s_{n+1}^1)$	Transition path and transmitted signal (i)
1	11	000	110	110	6
2	00	110	101	001	5
3	01	001	110	010	6
4	01	010	101	011	5
5	10	011	001	101	1
6	11	101	000	110	0

in the two paths. For instance, we calculate the distance between the two labeled paths in Fig. 4.7, namely $(0, 0, 0)$ and $(6, 7, 6)$, by first finding the distances between signals 0 and 6, 0 and 7, and 0 and 6, which are d_1, d_0, and d_1, respectively. Then the total distance between the two paths (or sequences) is $\sqrt{d_1^2 + d_0^2 + d_1^2} = 2.141$. Similarly, it can *be found that* the distance between the path sequence $(4, 1, 2)$ and $(6, 5, 6)$ is $\sqrt{d_1^2 + d_0^2 + d_2^2} = 2.566$.

The quantity called *free distance* and denoted by d_{free} is important, since the probability of an error event (deciding on a path diverging from the true path) for optimal soft decision decoding and an AWGN channel is lower bounded at high signal-to-noise ratios by

$$P[\text{ error event }] \geq K(d_{\text{free}})\left\{\frac{1}{2} - \text{erf}\left[\frac{d_{\text{frec}}}{2\sigma}\right]\right\}, \qquad (4.4.1)$$

where $K(d_{\text{free}})$ is the average number of error events at distance d_{free}. The calculation of d_{free} generally requires an exhaustive search over all possible error events, which we do not describe here. The comparison of two modulation/coding schemes involves the calculation of what is called the *coding gain,* defined as

$$\text{coding gain} \triangleq 10 \log_{10} \frac{\left(d_{\text{free}}^2 / P_{\text{av}}\right)_{\text{coded}}}{\left(d_{\text{min}}^2 / P_{\text{av}}\right)_{\text{uncoded}}}, \tag{4.4.2}$$

where d_{min} is the minimum distance between transmitted signal points in the uncoded constellation and P_{av} is the average power for each signal set. If we fix P_{av} to be the same for both coded and uncoded signal sets, as is usually done, we can set $P_{\text{av}} = 1$, so Eq. (4.4.2) becomes

$$\text{coding gain} \triangleq 10 \log_{10} \frac{\left(d_{\text{free}}^2\right)_{\text{coded}}}{\left(d_{\text{min}}^2\right)_{\text{uncoded}}}. \tag{4.4.3}$$

Thus we can compute the coding gain for the coded modulation method in Example 4.4.1 with respect to four-phase PSK by noting that for four-phase PSK $d_{\text{min}} = \sqrt{2}$, and since we know from Fig. 4.7 and our calculations that $d_{\text{free}} = 2.141$, then

$$\text{coding gain} = 10 \log_{10} \frac{(2.141)^2}{\left(\sqrt{2}\right)^2} = 3.6 \text{ dB}, \tag{4.4.4}$$

as specified in Fig. 4.7. The maximum practical improvement available is about 6 dB, although we cannot delve into this topic in this book.

Example 4.4.2 We consider the transmission of 3 bits/symbol using a 16-point QAM signal constellation with coding (Ungerboeck 1982). The partitioning of the 16-point QAM constellation into subsets with increasing Euclidean distance is demonstrated in Fig. 4.9. The convolutional encoder chosen and its associated trellis are shown in Figs. 4.10 and 4.11, respectively. Note that the rate-$\frac{3}{4}$ convolutional code in Fig. 4.10 is just the rate-$\frac{2}{3}$ convolutional code used in Example 4.4.1 with an additional uncoded bit. The coded bits select the specific subset, and the uncoded bit chooses the particular signal within that subset. Note also that the trellis diagram in Fig. 4.10 has double lines drawn between states. These double lines are called *parallel transitions* and are due to the uncoded bit being either 0 or 1.

For uncoded eight-phase PSK, $d_{\text{min}} = 0.765$, so the coding gain of the coded 16-point QAM scheme (with $d_{\text{free}} = \sqrt{2}$) over eight-phase PSK is 5.3 dB. If we compare the coded QAM method with the eight-point AM/PM scheme shown in Fig. 4.12, we find that

$$\text{coding gain} = 10 \log_{10} \frac{2}{4/5}$$

$$= 3.98 \text{ dB} \tag{4.4.5}$$

The problems investigate additional aspects of this example.

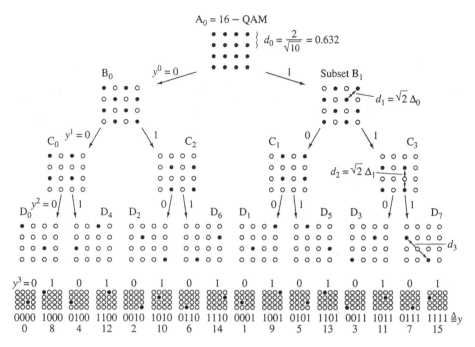

Fig. 4.9 Mapping by set partitioning of the 16-point QAM constellation. From G. Ungerboeck, "Channel Coding with Multilevel/Phase Signals," *IEEE Trans. Inf. Theory,* © 1982 IEEE

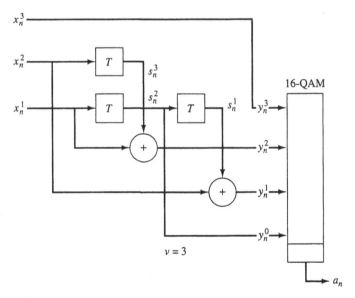

Fig. 4.10 Rate-$\frac{3}{4}$ convolutional encoder for Example 4.4.2. From G. Ungerboeck, "Channel Coding with Multilevel/Phase Signals," *IEEE Trans. Inf. Theory,* © 1982 IEEE

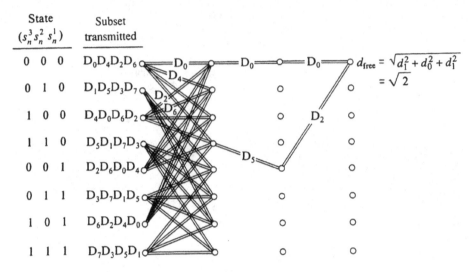

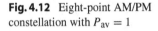

Fig. 4.11 Trellis diagram for convolutional code indicating allowable phase transitions for 3 bits/symbol in Example 4.4.2. From G. Ungerboeck, "Channel Coding with Multilevel/Phase Signals," *IEEE Trans. Inf. Theory*, © 1982 IEEE

Fig. 4.12 Eight-point AM/PM constellation with $P_{av} = 1$

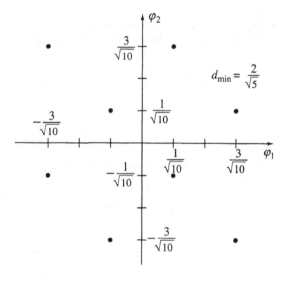

We have given heuristic arguments for how trellis-coded modulation provides a performance improvement, we have described two TCM systems in the examples and have demonstrated their operation, and we have introduced the concept of coding gain. It is beyond the scope of this book to develop the techniques for designing TCM systems and to provide the asymptotic theory that bounds TCM system performance. However, with the material in this section and in Chap. 3, the reader should be equipped to understand existing TCM systems and to begin to read the literature on the subject.

4.5 Summary

We have described in this chapter some of the earliest and at present, most successful systems that combine the operations of coding and modulation. Coded modulation is a dynamic and exciting field, and the reader is referred to more advanced texts and technical papers for a more complete discussion.

4.6 Problems

4.1 A transmitted signal $s(t)$ is a weighted linear combination of nonoverlapping time-shifted pulses $p(t)$ with

$$s(t) = \sum_{j=1}^{M} s_j p(t - jT),$$

where $p(t) \neq 0$ for $-T \leq t \leq 0$, $p(t) = 0$ for all other t, and

$$\int_{-T}^{0} p^2(t)dt = E_b$$

is the energy transmitted per bit. We assign the s_j according to

$$s_j = \begin{cases} +1, & \text{if the } j\text{th message bit is } 1 \\ -1, & \text{if the } j\text{th message bit is } 0. \end{cases}$$

If $s(t)$ is sent over an AWGN channel with power spectral density $\mathcal{N}_0/2$ W/Hz, show that $P[\mathcal{E}] = 1 - (1 - p)^M$, where $p = \frac{1}{2} - \text{erf}\left[\sqrt{2E_b/\mathcal{N}_0}\right]$. This has been called "bit-by-bit" signaling (Wozencraft and Jacobs 1965). The signal $s(t)$ can be represented geometrically as 2^M signal points at the vertices of a hypercube in M dimensions. What happens to $P[\mathcal{E}]$ as $M \to \infty$?

4.2 An M-bit binary message sequence is to be transmitted in a time interval τ seconds long. The transmitted signals are equally likely and of the form

$$s_i(t) = \sqrt{E}\varphi(t - iT),$$

where $\varphi(t) \neq 0$ for $0 \leq t \leq T, \varphi(t) = 0$ for t otherwise, $\int_0^T \varphi^2(t)dt = 1$, $T = \tau/2^M$, and $i = 0, 1, 2, \ldots, 2^M - 1$ is the number corresponding to the binary message sequence. Show that $P[\mathcal{E}] \leq (2^M - 1)p$, where $p = \frac{1}{2} - \text{erf}[\sqrt{E/N_0}]$, if these signals are transmitted over an AWGN channel with spectral density $N_0/2$ W/Hz. This has been called "block orthogonal" signaling (Wozencraft and Jacobs 1965). Note that the energy transmitted per bit $E_b = E/M$, and consider $P[\mathcal{E}]$ as $M \to \infty$. Compare to the result in Problem 4.1.

4.3   The pre-envelope $s(t)$ of a real bandpass signal $x(t)$ is defined as (Rice 1982)

$$s(t) = \hat{r}(t)e^{j\omega_0 t + j\hat{\varphi}(t)} = x(t) + jx_h(t),$$

where ω_0 is arbitrary and $x_h(t)$ is the Hilbert transform of $x(t)$. Clearly, $\text{Re}[s(t)] = x(t)$ and $\hat{r}(t) = \sqrt{x^2(t) + x_h^2(t)}$ is the magnitude of $s(t)$.

(a) Let $x(t) = m(t)\cos\omega_c t$ and show that $\hat{r}(t)$ is the envelope of $x(t)$

(b) Given a signal of the form

$$z(t) = x(t)\cos\omega_0 t - y(t)\sin\omega_0 t,$$

where the highest frequency in $x(t)$ and $y(t)$ is less than ω_0, find an expression for the analytic signal corresponding to $z(t)$ and expressions for $x(t)$ and $y(t)$.

4.4   Derive Eqs. (4.2.9) and (4.2.10).

4.5   Obtain $r(t)$ in Eq. (4.2.12) for the case where $-T \leq t \leq 0$.

4.6   Derive Eqs. (4.2.14) and (4.2.15).

4.7   Obtain the frequency pulses corresponding to the smoothing functions in Eqs. (4.3.5) and (4.3.6).

4.8   Sketch the smoothing function in Eq. (4.3.7) and its frequency pulse for $L = 2$ and $L = 3$.

4.9   Show that the minimum spacing of FSK frequencies for noncoherent orthogonality is the symbol rate, $1/T$.

4.10 Given a binary data sequence $\{b_k, k = 0, 1, 2, \ldots\}$ with $b_k = \pm 1$, a quadrature PSK (QPSK) signal can be written as

$$s(t) = \frac{b_i(t)}{\sqrt{2}}\cos\left[\omega_c t + \frac{\pi}{4}\right] + \frac{b_q(t)}{\sqrt{2}}\sin\left[\omega_c t + \frac{\pi}{4}\right],$$

where $b_i(t)$ is a rectangular pulse $2T$ seconds wide with an amplitude (+1) determined by the even-subscripted bits, $k = 0, 2, 4, \ldots$ Similarly, $b_q(t)$ is a rectangular pulse whose amplitude is specified by the odd-subscripted bits, $k = 1, 3, 5, \ldots$ (Pasupathy 1979).

(a) Show that the possible transmitted phases are 0; $\pm 90°$; and 180° every $2T$ seconds.

(b) For the input binary sequence $\{b_k\} = \{1, -1, 1, -1, -1, -1, 1, 1\}$, specify the transmitted phase sequence and sketch the transmitted waveform.

4.11 Offset QPSK (OQPSK), also called staggered QPSK, can be generated from the formulation in Problem 4.10 by delaying $b_q(t)$ with respect to $b_i(t)$ by T seconds (Pasupathy 1979).

(a) Show that the possible transmitted phases are 0 and $\pm 90°$ every T seconds.

(b) For the same binary input sequence as in Problem 4.10(b), find the transmitted phase sequence and sketch the transmitted waveform for OQPSK.

(c) Compare part (b) of Problems 4.10 and 4.11.

4.12 MSK can be written as a version of OQPSK described in Problem 4.11 in which the quadrature components have cosine and sine pulse shaping.

In particular, an MSK waveform can be expressed as

$$s(t) = b_i(t) \cos \frac{\pi t}{2T} \cos \omega_c t + b_q(t) \sin \frac{\pi t}{2T} \sin \omega_c t,$$

where $b_i(t)$ and $b_q(t)$ are as defined in Problems 4.10 and 4.11. (a) Show that $s(t)$ can be rewritten as

$$s(t) = \cos \left[\omega_c t - b_i(t) b_q(t) \frac{\pi t}{2T} + \phi_k \right],$$

where

$$\phi_k = \begin{cases} 0, & \text{if } b_i(t) = +1 \\ \pi, & \text{if } b_i(t) = -1. \end{cases}$$

(b) Use $s(t)$ in part (a) to demonstrate that MSK is also an FSK signal transmitting the two frequencies $\omega_+ = \omega_c + \pi/2T$ and $\omega_- = \omega_c - \pi/2T$ (Pasupathy 1979).

4.13 For the binary input message sequence $\{+1, +1, +1, +1, -1, -1, +1, -1\}$, find the transmitted phases at the time instants $nT, n \geq 1$ for MSK using the trellis diagram in Fig. 4.3.

4.14 Draw a trellis diagram corresponding to the phase smoothing function $f(t)$ in Fig. 4.2c. Let $h = \frac{1}{2}$ and $M = 2$.

4.15 Draw a phase tree for multi-h CPM for $f(t)$ in Fig. 4.2b, $M = 2$, and $\{h_1, h_2\} = \left\{\frac{3}{8}, \frac{4}{8}\right\}$ (Ziemer and Peterson 1985).

4.16 Find the sequence of transmitted phases for the multi-h CPM system in Problem 4.15 if the input binary message string is

$$\{+1, -1, -1, -1, +1, +1, +1, -1\}.$$

4.17 For the coded modulation scheme in Example 4.4.1, let the initial encoder state be $\left(s_0^3 \ s_0^2 \ s_0^1 \right) = (0\,0\,0)$ and assume that the input sequence is 101101110101. Specify the current state, coder output, path transition, transmitted symbol, and next state as a function of time n.

4.18 Find the distance between the path $(0, 0, 0)$ and $(6, 5, 2)$ in Problem 4.17.

4.19 Three eight-phase PSK coded modulation methods for sending 2 bits/symbol have the d_{free} shown

$$\text{Code:} \quad 1 \quad 2 \quad 3$$
$$d_{\text{free}} : \quad 1.608 \quad 2.0 \quad 2.274$$

Calculate the coding gain of each of these techniques with respect to four-phase PSK.

4.20 A rate-$\frac{2}{3}$ convolutional coder is shown in the below figure (a) along with its associated trellis in below figure (b). A bit-to-symbol mapping for a one-dimensional signal set is shown in below figure (c). For this coded modulation method, $d_{\text{min}}^2 / P_{\text{av}} = 36/21$ (Thapar 1984).

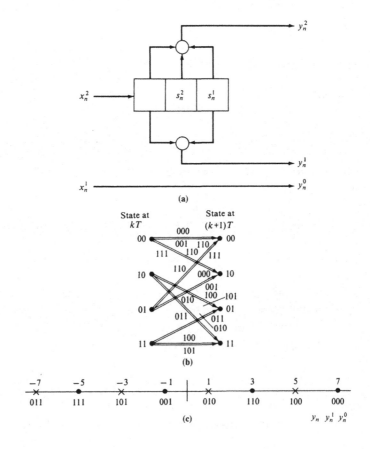

(a) For an initial encoder state of $\left(s_0^2 \, s_0^1\right) = (0\,0)$ and the input sequence 11011010, find the current state, coder output, path transition, transmitted symbol, and next state as a function of time.

(b) Find the coding gain with respect to a four-level one-dimensional signal set.

4.21 For the trellis in above figure(b), find the distance between the paths (7, 7, 7) and (3, −3, −1).

4.22 Redraw the one-dimensional mapping in above figure (c) as a circle, thus generating a two-dimensional mapping. Calculate the coding gain of this new code with respect to four-phase PSK (Thapar 1984).

4.23 For the coded modulation system in Example 4.4.2, let the initial state be $\left(s_n^3 \, s_n^2 \, s_n^1\right) = (0\,0\,0)$ and assume that the input message sequence is 101110111001. Find the current state, coder output, path transition, transmitted symbol, and next state as a function of time n.

4.24 Given the coded modulation method in Example 4.4.2, calculate the distance between the two paths (8, 6, 14) and (12, 14, 9).

4.25 Parallel transitions allow single-bit-error events to occur, which implies that d_{free} is less than or equal to d_{min} for the signal subset corresponding to the parallel transition. What is d_{min} for the parallel transition subsets in Fig. 4.11?

4.26 Find the distance between parallel transitions for the code in the figure above.

4.27 Draw a one-state trellis diagram for uncoded four-phase PSK at a rate of 2 bits/symbol. Specify d_{free}.

References

Abramson, N. 1973. "The Aloha System." Chapter 14 in *Computer Communication Networks.* N. Abramson and F. F. Kuo, eds. Englewood Cliffs, N.J.: Prentice Hall, pp. 501–517.

Amoroso, F. 1980. "The Bandwidth of Digital Data Signals." *IEEE Commun. Mag.,* Vol. 18, Nov.

Anderson, J. B., and J. R. Lesh, eds. 1981. "Special Section on Combined Coding and Modulation." *IEEE Trans. Commun.,* Vol. COM-29, Mar.

Anderson, J. B, C.-E. W. Sundberg, T. Aulin, and N. Rydbeck. 1981. "Power-Bandwidth Performance of Smoothed Phase Modulation Codes," *IEEE Trans. Commun.,* Vol. COM-29, Mar., pp. 187–195.

Anderson, R. R., and J. Salz. 1965. "Spectra of Digital FM." *Bell Syst. Tech. J.,* Vol. 44, July–Aug., pp. 1165–1189.

AT&T, *Telecommunications Transmission Engineering,* Vols. 1–3. New York: AT&T, 1977.

Bellamy, J. 1982. *Digital Telephony.* New York: Wiley.

Berger, T. 1971. *Rate Distortion Theory: A Mathematical Basis for Data Compression.* Englewood Cliffs, N.J.: Prentice Hall.

Bertsekas, D., and R. Gallager. 1987. *Data Networks.* Englewood Cliffs, N.J.: Prentice Hall.

Bhargava, V. K., D. Haccoun, R. Matyas, and P. Nuspl. 1981. *Digital Communications by Satellite.* New York: Wiley.

Blahut, R. E. 1983. *Theory and Practice of Error Control Codes.* Reading, Mass.: Addison-Wesley.

Blahut, R. E. 1990. *Digital Transmission of Information.* Reading, Mass.: Addison-Wesley.

Briley, B. E. 1983. *Introduction to Telephone Switching.* Reading, Mass.: Addison-Wesley.

Capon, J. 1959. "A Probabilistic Model for Run-Length Coding of Pictures." *IRE Trans. Inf. Theory,* Vol. IT-5, pp. 157–163.

Carlson, A. B. 1975. *Communication Systems.* New York: McGraw-Hill.

Chen, W. H. 1963. *The Analysis of Linear Systems.* New York: McGraw-Hill.

Clark, G. C., Jr., and J. B. Cain. 1981. *Error-Correction Coding for Digital Communications.* New York: Plenum Press.

Clos, C. 1953. "A Study of Non-blocking Switching Networks," *Bell Syst. Tech. J.,* Vol. 32, Mar., pp. 406–424.

Collins, A. A., and R. D. Pedersen. 1973. *Telecommunications: A Time for Innovation.* Dallas, Tex.: Merle Collins Foundation.

Cooper, G. R., and C. D. McGillem. 1986. *Modem Communications and Spread Spectrum.* New York: McGraw-Hill.

Davies, D. W., and D. L. A. Barber. 1973. *Communication Networks for Computers.* New York: Wiley.

© The Editor(s) (if applicable) and The Author(s), under exclusive license to Springer Nature Switzerland AG 2023

J. D. Gibson, *Digital Communications,* Synthesis Lectures on Communications, https://doi.org/10.1007/978-3-031-19588-4

de Buda, R. 1972. "Coherent Demodulation of Frequency Shift Keying with Low Deviation Ratio." *IEEE Trans. Commim.*, Vol. COM-20, June, pp. 429–436.

Dixon, R. C. 1976. *Spread Spectrum Techniques.* New York: IEEE Press.

Dixon, R. C. 1984. *Spread Spectrum Systems.* New York: Wiley.

Farvardin, N., and J. W. Modestino. 1984. "Optimum Quantizer Performance for a Class of Non-Gaussian Memoryless Sources." *IEEE Trans. Inf. Theory,* Vol. IT-30, May, pp. 485–497.

Forney, G. David, Jr. 1989. "Introduction to Modem Technology: Theory and Practice of Bandwidth Efficient Modulation from Shannon and Nyquist to Date." University Video Communications and the IEEE; Copyright: Motorola.

Foschini, G. J., R. D. Gitlin, and S. B. Weinstein. 1974. "Optimization of Two-Dimensional Signal Constellations in the Presence of Gaussian Noise." *IEEE Trans. Commim.,* Vol. COM-22, Jan., pp. 28–38.

Franks, L. E. 1969. *Signal Theory.* Englewood Cliffs, N. J.: Prentice Hall.

Franks, L. E. 1980. "Carrier and Bit Synchronization in Data Communication: A Tutorial Review." *IEEE Trans. Commun.,* Vol. COM-28, Aug., pp. 1107–1121.

Freeman, R. L. 1981. *Telecommunications Transmission Handbook,* 2nd ed. New York: Wiley.

Gallager, R. G. 1968. *Information Theory and Reliable Communication.* New York: Wiley.

Gersho, A., and R. M. Gray. 1991. *Vector Quantization and Signal Compression.* Hingham, Mass.: Kluwer.

Gilhousen, K. S., I. M. Jacobs, R. Padovani, and L. A. Weaver, Jr. 1990. "Increased Capacity Using CDMA for Mobile Satellite Communication." *IEEE J. Sel. Areas Commun.,* Vol. SAC-8, May, pp. 503–514.

Gold, R. 1967. "Optimal Binary Sequences for Spread-Spectrum Multiplexing." *IEEE Trans. Inf. Theory,* Vol. IT-13, pp. 619–621.

Gold, R. 1968. "Maximal Recursive Sequences with 3-Valued Recursive Cross Correlation Functions." *IEEE Trans. Inf. Theory,* Vol. IT-14, Jan., pp. 154–156.

Golomb, S. W., ed. 1964. *Digital Communications with Space Applications.* Englewood Cliffs, N.J.: Prentice Hall.

Gray, R. M., and L. D. Davisson. 1986. *Random Processes: A Mathematical Approach for Engineers.* Englewood Cliffs, N.J.: Prentice Hall.

Hammond, J. L., and P. J. P. O'Reilly. 1986. *Performance Analysis of Local Computer Networks.* Reading, Mass.: Addison-Wesley.

Haykin, S. 1983. *Communication Systems.* New York: Wiley.

Heller, J. A., and I. M. Jacobs. 1971. "Viterbi Decoding for Satellite and Space Communications." *IEEE Trans. Commun. Technol,* Vol. COM-19, Oct., pp. 835–848.

Hirsch, D., and W. J. Wolf. 1970. "A Simple Adaptive Equalizer for Efficient Data Transmission." *IEEE Trans. Commun. Technol.,* Vol. COM-18, Feb., pp. 5–12.

Holmes, J. K. 1982. *Coherent Spread Spectrum Systems.* New York: Wiley.

Holzman, L. N, and W. J. Lawless. 1970. "Data Set 203: A New High-Speed Voiceband Modem." *Computer,* Sept.–Oct., pp. 24–30.

Houston, S. W. 1975. "Modulation Techniques for Communication, Part I: Tone and Noise Jamming Performance for Spread Spectrum *M*-ary FSK and 2, 4-ary DPSK Waveforms." Proceedings of the IEEE National Aerospace and Electronics Conference (NAECON'75), Dayton, Ohio, June 10–12, pp. 51–58.

Huffman, D. A. 1952. "A Method for the Construction of Minimum Redundancy Codes." *Proc. IRE,* Vol. 40, Sept., pp. 1098–1101.

Jackson, D. 1941. *Fourier Series and Orthogonal Polynomials.* Washington, D.C.: The Mathematical Association of America.

Jacobaeus, C. 1950. "A Study of Congestion in Link Systems." *Ericsson Tech.,* No. 48, Stockholm.

Jahnke, E., and F. Emde. 1945. *Tables of Functions*. New York: Dover.

Jayant, N. S., and P. Noll. 1984. *Digital Coding of Waveforms*. Englewood Cliffs, N.J.: Prentice Hall.

Johnson, G. D. 1973. "No. 4 ESS." *Bell Lab. Rec.,* Sept., pp. 226–232.

Kaplan, W. 1959. *Advanced Calculus*. Reading, Mass.: Addison-Wesley.

Keiser, G. E. 1989. *Local Area Networks*. New York: McGraw-Hill.

Kernighan, B. W., and S. Lin. 1973. "Heuristic Solution of a Signal Design Optimization Problem." *Proc. 7th Annual Princeton Conference on Information Science and Systems,* Mar.

Kleinrock, L., and S. S. Lam. 1975. "Packet Switching in a Multiaccess Broadcast Channel: Performance Evaluation." *IEEE Trans. Commun.,* Vol. COM-23, Apr., pp. 410–423.

Kotel'nikov, V. A. 1947. *The Theory of Optimum Noise Immunity*. Doctoral dissertation, Molotov Energy Institute, Moscow. Also published by McGraw-Hill, New York, 1959.

Kretzmer, E. R. 1965. "Binary Data Communication by Partial Response Transmission." *Conf. Rec.,* 1965 IEEE Annual Communications Conference, pp. 451–455.

Kretzmer, E. R. 1966. "Generalization of a Technique for Binary Data Communication." *IEEE Trans. Commun. Technol.,* Feb., pp. 67–68.

Kuo, F. F. 1962. *Network Analysis and Synthesis*. New York: Wiley.

Lathi, B. P. 1968. *Communication Systems*. New York: Wiley.

Lee, C. Y. 1955. "Analysis of Switching Networks." *Bell Syst. Tech. J.,* Vol. 34, Nov., pp. 1287–1315.

Lender, A. 1963. "The Duobinary Technique for High Speed Data Transmission." *IEEE Trans. Commun. Electron.,* Vol. 82, May, pp. 214–218.

Lender, A. 1964. "Correlative Digital Communication Techniques." *IEEE Trans. Commun. Technol,* Dec., pp. 128–135.

Lender, A. 1966. "Correlative Level Coding for Binary Data Transmission." *IEEE Spectrum,* Vol. 3, Feb., pp. 104–115.

Lender, A. 1981. Chapter 7 in *Digital Communications: Microwave Applications*. K. Feher, ed. Englewood Cliffs, N.J.: Prentice Hall, pp. 144–182.

Levitt, B. K. 1985. "Strategies for FH/MFSK Signaling with Diversity in Worst-Case Partial Band Noise." *IEEE J. Sel. Areas Commun.,* Vol. SAC-3, Sept., pp. 622–626.

Lin, S., and D. J. Costello, Jr. 1983. *Error Control Coding: Fundamentals and Applications*. Englewood Cliffs, N.J.: Prentice Hall.

Lloyd, S. P. 1982. "Least Squares Quantization in PCM." *IEEE Trans. Inf. Theory,* Vol. IT-28, Mar., pp. 129–137 (unpublished memorandum, Bell Laboratories, 1957).

Lucky, R. W. 1965. "Automatic Equalization for Digital Communications." *Bell Syst. Tech. J.,* Vol. 44, Apr., pp. 547–588.

Lucky, R. W. 1966. "Techniques for Adaptive Equalization of Digital Communication." *Bell Syst. Tech. J.,* Vol. 45, Feb., pp. 255–286.

Lucky, R. W., and H. Rudin. 1967. "An Automatic Equalizer for General-Purpose Communication Channels." *Bell Syst. Tech. J.,* Vol. 46, Nov., pp. 2179–2207.

Lucky, R. W., J. Salz, and E. J. Weldon, Jr. 1968. *Principles of Data Communication*. New York: McGraw-Hill.

Makhoul, J. 1975. "Linear Prediction: A Tutorial Review." *Proc. IEEE,* Vol. 63, Apr., pp. 561–580.

Martin, D. R., and P. L. McAdam. 1980. "Convolutional Code Performance with Optimal Jamming." *Conf. Rec.,* 1980 IEEE International Conference on Communications, Seattle, Wash., June 8–12, pp. 4.3.1–4.3.7.

Max, J. 1960. "Quantizing for Minimum Distortion." *IRE Trans. Inf. Theory,* Vol. IT-6, Mar., pp. 7–12.

Mazur, B. A., and D. P. Taylor. 1981. "Demodulation and Carrier Synchronization of Multi-h Phase Codes." *IEEE Trans. Commun.,* Vol. COM-29, Mar., pp. 257–266.

McEliece, R. J. 1977. *The Theory of Information and Coding*. Reading, Mass.: Addison-Wesley.

Newman, D. B., Jr., and R. L. Pickholtz, eds. 1987. Special Issue on "Network Security." *IEEE Network,* Vol. 1, Apr.

Noll, P., and R. Zelinski. 1978. "Bounds on Quantizer Performance in the Low Bit-Rate Region." *IEEE Trans. Commun.,* Vol. COM-26, Feb., pp. 300–304.

Nyquist, H. 1924. "Certain Factors Affecting Telegraph Speed." *Bell Syst. Tech. J.,* Vol. 3, Apr., pp. 324–346.

Nyquist, H. 1928. "Certain Topics in Telegraph Transmission Theory." *Trans. AIEE,* Vol. 47, Apr, pp. 617–644.

Owen, F. F. E. 1982. *PCM and Digital Transmission Systems.* New York: McGraw-Hill.

Paez, M. D., and T. H. Glisson. 1972. "Minimum Mean-Squared-Error Quantization in Speech PCM and DPCM Systems." *IEEE Trans. Commun.,* Vol. COM-20, Apr., pp. 225–230.

Pahlavan, K, and J. L. Holsinger. 1988. "Voice-Band Data Communication Modems: A Historical Review: 1919–1988." *IEEE Commun. Mag.,* Vol. 26, Jan, pp. 16–27.

Pasupathy, S. 1977. "Correlative Coding: A Bandwidth-Efficient Signaling Scheme." *IEEE Commun. Mag.,* Vol. 15, July, pp. 4–11.

Pasupathy, S. 1979. "Minimum Shift Keying: A Spectrally Efficient Modulation." *IEEE Commun. Mag.,* Vol. 17, July, pp. 14–22.

Pickholtz, R. L, D. L. Schilling, and L. B. Milstein. 1982. "Theory of Spread-Spectrum Communications: A Tutorial." *IEEE Trans. Commun.,* Vol. COM-30, May, pp. 855–884.

Pickholtz, R. L, D. L. Schilling, and L. B. Milstein. 1984. "Revisions to 'Theory of Spread Spectrum Communications: A Tutorial'," *IEEE Trans. Commun.,* Vol. COM-32, Feb, pp. 211–212.

Pratt, W. K. 1978. *Digital Image Processing.* New York: Wiley.

Proakis, J. G. 1989. *Digital Communications.* New York: McGraw-Hill.

Qureshi, S. U. H. 1985. "Adaptive Equalization." *Proc. IEEE,* Vol. 73, Sept., pp. 1349–1387.

Rabiner, L. R., and R. W. Schafer. 1978. *Digital Processing of Speech Signals.* Englewood Cliffs, N.J.: Prentice Hall.

Rice, S. O. 1963. "Noise in FM Receivers." Chapter 25 in *Proc., Symposium on Time Series Analysis.* M. Rosenblatt, ed. New York: Wiley, pp. 395–424.

Rice, S. O. 1982. "Envelopes of Narrow-Band Signals." *Proc. IEEE,* Vol. 70, July, pp. 692–699.

Sarwate, D. V., and M. B. Pursley. 1980. "Crosscorrelation Properties of Pseudorandom and Related Sequences." *Proc. IEEE,* Vol. 68, May, pp. 593–619.

Schilling, D. L., L. B. Milstein, R. L. Pickholtz, and R. W. Brown. 1980. "Optimization of the Processing Gain of an *M*-ary Direct Sequence Spread Spectrum Communication System." *IEEE Trans. Commun.,* Vol. COM-28, Aug., pp. 1389–1398.

Schilling, D. L., R. L. Pickholtz, and L. B. Milstein, guest eds. 1990. "Spread Spectrum Communications I." Special issue, *IEEE J. Sel. Areas Commun.,* Vol. SAC-8, May.

Scholtz, R. A. 1982. "The Origins of Spread-Spectrum Communications." *IEEE Trans. Commun.,* Vol. COM-30, May, pp. 822–854.

Schwartz, L. 1950. *Theorie des Distributions,* Vol. 1. Paris: Hermann.

Schwartz, M. 1987. *Telecommunications Networks: Protocols, Modeling, and Analysis.* Reading, Mass.: Addison-Wesley.

Schwartz, M. 1990. *Information Transmission, Modulation, and Noise,* 4th ed. New York: McGraw-Hill.

Shannon, C. E. 1948. "A Mathematical Theory of Communication." *Bell Syst. Tech. J.,* Vol. 27, July, pp. 379–423; Oct., pp. 623–656.

Shannon, C. E. 1959. "Coding Theorems for a Discrete Source with a Fidelity Criterion." *IRE Natl. Conv. Rec.,* Pt. 4, Mar., pp. 142–163.

Shannon, C. E., and W. Weaver. 1949. *The Mathematical Theory of Communication.* Urbana, Ill.: University of Illinois Press.

Simon, M. K., J. K. Omura, R. A. Scholtz, and B. K. Levitt. 1985. *Spread Spectrum Communications,* Vols. I and II, Rockville, Md.: Computer Science Press.

Sklar, B. 1988. *Digital Communications: Fundamentals and Applications.* Englewood Cliffs, N.J.: Prentice Hall.

Spragins, J. D., J. L. Hammond, and K. Pawlikowski. 1991. *Telecommunications: Protocols and Design.* Reading, Mass.: Addison-Wesley.

Stallings, W. 1984. "Local Network Performance." *IEEE Commun. Mag.,* Vol. 22, Feb., pp. 27–36.

Sunde, E. D. 1961. "Pulse Transmission by AM, FM, and PM in the Presence of Phase Distortion." *Bell Syst. Tech. J.,* Vol. 40, Mar., pp. 353–422.

Tanenbaum, A. S. 1988. *Computer Networks.* Englewood Cliffs, N.J.: Prentice Hall.

Taub, H., and D. L. Schilling. 1986. *Principles of Communication Systems, 2nd ed.* New York: McGraw-Hill.

Temple, G. 1953. "Theories and Applications of Generalized Functions." *J. London Math. Soc.,* Vol. 28, pp. 134–148.

Thapar, H. K. 1984. "Real-Time Application of Trellis Coding to High-Speed Voiceband Data Transmission." *IEEE J. Sel. Areas Commun.,* Vol. SAC-2, Sept., pp. 648–658.

Thomas, G. B., Jr. 1968. *Calculus and Analytic Geometry.* Reading, Mass.: Addison-Wesley.

Ungerboeck, G. 1982. "Channel Coding with Multilevel/Phase Signals." *IEEE Trans. Inf. Theory,* Vol. IT-28, Jan., pp. 55–67.

Viswanathan, R., and K. Taghizadeh. 1988. "Diversity Combining in FH/BFSK Systems to Combat Partial Band Jamming." *IEEE Trans. Commun.,* Vol. COM-36, Sept., pp. 1062–1069.

Viterbi, A. J., and I. M. Jacobs. 1975. "Advances in Coding and Modulation for Noncoherent Channels Affected by Fading, Partial Band, and Multiple Access Interference." In *Advances in Communication Systems,* Vol. 4. A. V. Balakrishnan, ed. New York: Academic Press.

Wozencraft, J. M., and I. M. Jacobs. 1965. *Principles of Communication Engineering.* New York: Wiley.

Wyner, A. D. 1981. "Fundamental Limits in Information Theory." *Proc. IEEE,* Vol. 69, Feb., pp. 239–251.

Ziemer, R. E., and R. L. Peterson. 1985. *Digital Communications and Spread Spectrum Systems.* New York: Macmillan.

Printed in the United States
by Baker & Taylor Publisher Services